# PRINCIPLES

## OF

# ENVIRONMENTAL,

# HEALTH

# SAFETY

# MANAGEMENT

*Editors*

**Gordon A. West**
**Ronald W. Michaud**

T0290671

**Government Institutes, Inc.**
**Rockville, Maryland**

Government Institutes, Inc., 4 Research Place, Suite 200
Rockville, Maryland 20850

99   98                          5

Printed in the United States of America.

*Library of Congress Cataloging-in-Publication Data*

   Principles of environmental health and safety managment / editors Gordon
A. West, Ronald W. Michaud.
   p.    cm.
Includes bibliographical references and index.
ISBN:  0-86587-478-6
   1. Industrial management--Environmental aspects   2. Industrial hygiene--
Managment    I. West, Gordon A.
II. Michaud, Ronald W.
HD30.255.P75    1995
658.4'08--dc20

                                        95-31972 CIP

# PRINCIPLES OF ENVIRONMENTAL, HEALTH AND SAFETY MANAGEMENT

## Table of Contents

# PUBLISHER'S ACKNOWLEDGEMENT

I would like to extend my appreciation to all of the authors who have taken time from their busy schedules to contribute to this new edition, and especially to Gordon West and Ron Michaud who oversaw the development of the project from start to finish. Without their efforts, this book would not have been possible.

I sincerely hope that readers will find this to be a useful reference. I encourage all those who want additional information in a given area to see the many references listed in the back of this handbook.

As always, we welcome the opportunity to hear from our readers as to how well this publication, as well as all of our products, continue to help meet our mutual goals of compliance with the "spirit" as well as the "letter" of environmental, health and safety regulations.

Thomas F.P. Sullivan
President, Government Institutes, Inc.
July, 1995

# PREFACE

The purpose of this book is to provide information and advice on how a company can meet its obligations to its shareholders and to the environment in which it operates. The book is intended to be an aid in the training and development of environmental managers and of the people for whom and with whom they work. Environmental management is not the sole province of the environmental manager. It is the responsibility of the entire enterprise and cannot be met unless its demands and objectives are understood by all those whose efforts are required to meet them.

The chapters of this book have been written by persons with hands-on experience in both general and specific areas of environmental health and safety management. Their contributions have been made with the hope that their experiences will benefit others.

Gordon A. West.
Ronald W. Michaud

July, 1995

# About The Authors

**Braden R. Allenby**

Braden R. Allenby is currently the research vice president of technology and environment, for AT&T. He joined AT&T in 1983 as a telecommunications regulatory attorney, and was an environmental attorney and senior environmental attorney for AT&T from 1984 to 1993. During 1992, he was the J. Herbert Holloman Fellow at the National Academy of Engineering in Washington, D.C. He is currently the vice-chair of the IEEE Committee on the Environment; a member of the DOE Task Force on Alternative Futures for the DOE National Laboratories; a member of the National Research Council Committee on Research and Peer Review in EPA, and a member of the Advisory Committee of the UNEP Working Group on Product Design for Sustainability. Dr. Allenby graduated *cum laude* from Yale University, and received his J.D. from the University of Virginia Law School and his masters in economics from the University of Virginia. He received his masters in environmental sciences from Rutgers University and his Ph.D. in environmental sciences from Rutgers. He is a member of the Virginia Bar and has worked as an attorney for the Civil Aeronautics Board and the Federal Communications Commission, as well as a strategic consultant on economic and technical telecommunications issues. He has also authored a number of articles and books on industrial ecology and design for the environment, and teaches a course on industrial ecology at the Yale

University School of Forestry and Environmental Studies, an engineering extension course on design for the environment at the University of Wisconsin, and has lectured at a number of universities, including Dartmouth College, Harvard, M.I.T., Princeton, Rutgers, and Tufts. He is a Fellow of the Royal Society for the Arts, Manufacturers & Commerce.

## Maria M. Bober Rasmussen

Maria M. Bober Rasmussen has spent her entire career working for Kodak in the environmental area. Her assignments have included environmental services representative to the Chemicals & Recovery Division; RCRA compliance coordinator; leader of the air regulatory group (during the reauthorization of the Clean Air Act), and corporate environmental auditor. Prior to this assignment, she was on loan to Corporate Communications as Kodak's first coordinator of environmental communications. In that role, she became involved in the Global Environmental Management Initiative (GEMI), and became chair of GEMI's Environmental Reporting and Communications Workgroup. Recently, she returned to Corporate Health, Safety and Environment, where she is an issues manager, reporting to Kodak's vice-president of Health, Safety and Environment. Her areas of responsibility include energy; sustainable development; sourcing issues; coordination of HSE benchmarking, and contributions. Additionally, she became vice-chair of GEMI in June, 1995. She earned her B.S. degrees in chemical engineering and public policy from Carnegie Mellon University.

## David R. Chittick

David R. Chittick died shortly after the first publication of this book. He had recently retired from AT&T, where he served as vice-president of environment & safety engineering. In that position he was responsible for environmental, product, and occupational safety. Mr. Chittack was a founding member and chairman of the Environmental Management Roundtable, a group of industry executives who meet to discuss and share common environmental and safety problems and solutions. He was also a member of the U.S. Mission to the Peoples Republic of China regarding Stratospheric Ozone Depletion, and a member of a U.S. government delegation to the U.S.S.R. and Hungary, to exchange technical information regarding CFC solvent replacement. He received his B.S. in electrical engineering from the University of Vermont, and attended the Massachusetts Institute of Technology as a Sloan Fellow where he earned his M.S. degree. He was a frequent speaker at Government Institutes, American Electronics Association, and other association meetings, and was a member of several business and industry environmental coalitions. He made many important contributions to the field of environmental management.

## John W. Coryell

John W. Coryell currently serves as senior director of Environmental Affairs for Conoco, Inc., petroleum subsidiary of DuPont Company. He and his staff provide environmental policy guidance and business processes in support of Conoco's global businesses and they coordinate on these topics with the parent corporation. He has held a number of technical, marketing, and manufacturing assignments at DuPont, including that of environmental coordinator for their textile fibers businesses. He also served for a time, as the technical liason for DuPont's environmental management business in

Taiwan; he has been with Conoco for the last six years. Previous to his current position, he provided internal consultation on air compliance issues, PCBs, and SARA/CERCLA release reporting. He received his Ph.D. in physical chemistry from Oklahoma State University.

## Dana M. Glorie

Dana M. Glorie is a senior consultant in the metropolitan New York practice of Arthur Andersen. Over the last four years, she has gained extensive experience supervising project teams engaged in designing and implementing activity-based cost management models in the telecommunications, pharmaceuticals and industrial products industries. She also has significant experience in performing process improvement and benchmarking engagements.

## Michael D. Henke

Michael D. Henke has managed worldwide due diligence efforts on some of the largest acquisitions and divestitures in the chemical and energy industries for Pilko & Associates' industrial, financial, and legal clients. For several years, he coordinated Pilko & Associates' worldwide auditing and risk assessment activity. He earned a national reputation for his leading-edge work in environmental auditing, including development efforts in environmental management systems. He has developed compliance assurance and audit oversight programs, and has conducted Quality Assurance (QA) reviews of the audit programs of several Fortune 100 companies. He has also managed self-audit verification programs, participated on audit teams, conducted auditor training sessions, and facilitated environmental audit program benchmarking studies involving world-class companies from various industries. He also has broad experience in manufacturing and environmental affairs management. With Gulf Oil Chemicals (GOCHEM), he completed several assignments in plant operations and engineering. Subsequently, at GOCHEM headquarters, he instituted and conducted the company's environmental audit program. This program covered 70 manufacturing facilities and laboratories throughout the United States. Gulf Oil Corporation adopted the GOCHEM audit for use throughout the company. Later, as Coordinator for Energy Conservation, he was responsible for coordination of energy conservation, toxicology testing, and stock loss control programs for GOCHEM's manufacturing plants. He also helped coordinate the corporate response to environmental emergencies. He received his B.S. in engineering science from the University of Notre Dame, and his M.B.A. from the University of St. Thomas. He is currently active in the Environmental Auditing Roundtable.

## Robert B. McKinstry, Jr.

Robert B. McKinstry, Jr. is a partner in the environmental group of Ballard Spahr Andrews & Ingersoll in the firm's Philadelphia office. His practice focuses on all aspects of environmental law and litigation. He has written and spoken extensively on environmental topics before professional and industry groups. Mr. McKinstry received his B.A. from Swarthmore College with honors, an M.F.S. from the Yale School of Forestry and Environmental Studies, and a J.D. from the Yale Law School.

## Ronald W. Michaud

Ronald W. Michaud has over 28 years of industrial experience, including over nine years of management experience in the safety, health and environmental arena. He has been a leader in the areas of strategic environmental planning and development of proactive environmental programs and is a regular speaker at international, national, and regional environmental conferences. As the former corporate environmental stewardship leader for E.I. duPont de Nemours & Co., Inc., he was responsible for the corporate environmental organization charged with developing worldwide environmental policy; leading the environmental strategic planning process and program development; providing oversight function for the environmental audit program; establishing and leading networks for compliance guidance and implementation; managing corporate environmental philanthropy; initiating benchmarking activities, and coordinating environmental engineering activities. Mr. Michaud was also formerly the Corporate Environmental Manager for Conoco, Dupont's energy subsidiary, and prior to that, he was the Safety, Health and environmental affairs manager for Dupont's Fibers Department. In addition to his EH&S management background, he has broad experience in manufacturing, R&D, quality control, process development, and organizational effectiveness and people development. His broad industry and environmental management background is ideally suited to assist Pilko & Associates' clients with EH&S strategic planning, benchmarking, developing proactive EH&S approaches or organizational development. Mr. Michaud has served on the Advisory Board of the Harvard Center for Risk Analysis, the Board of Directors of the Institute for Cooperation in Environmental Management (ICEM), the American Petroleum Institute's Environment General Committee, and has for the last seven years been a member of the Speakers Bureau for Government Institute's Environmental Management Roundtable. He is also currently serving as Chairman of the Board of Trustees for the Nature Conservancy in Delaware. He received his B.S. in chemistry from Northeastern University in Boston, Massachusetts.

## George J. Miller

After service in the U.S. Army as appellate government counsel in the Office of the Judge Advocate General in the Pentagon, Mr. Miller engaged in private practice with Dechert Price and Rhoads as a trial and environmental partner. He was chair of that firm's environmental practice group for many years. He is a past member of the American Bar Association's Standing Committee of Environmental Law and is the principal author of a *Guide to Air Permitting and Enforcement*, published by M. Lee Smith Publishers and Printers in 1995. He was appointed to the Environmental Hearing Board and named its chairman by Governor Tom Ridge in May, 1995. He also recently started as chairman of Pennsylvania's Environmental Hearing Board. Mr. Miller received his AB degree from Princeton University and his JD from the University of Pennsylvania Law School. He is a member of the Philadelphia Bar Association and resides in Haverford, Pennsylvania.

## F. Paul Pizzi

Paul Pizzi is the senior vice-president of Pilko & Associates, Inc. He co-founded the firm in 1980, and is responsible for Pilko's worldwide environmental, health, and safety (EH&S) auditing business. He is currently leading Pilko's benchmarking activities and assisting leading companies in improving their EH&S performance. Mr. Pizzi has been actively involved in EH&S audits and risk assessments at both the corporate and facility level since 1975. He has helped many companies structure and implement EH&S audit and compliance review programs. He has also directed worldwide environmental due diligence assessments of businesses for acquisition and divestiture analysis. He started his professional career with Sinclair Refining Company and later continued with SOHIO and BP. He received his B.S. in chemical engineering from Villanova University, and is a frequent speaker to senior executives on strategic environmental planning in the 1990s, cooperative benchmarking, and trends in EH&S auditing.

## J. Richard Pooler

J. Richard Pooler is an Associate with Blasland, Bouck & Lee, Inc., an environmental consulting firm of over 500 professionals nationwide. He possesses more than 15 years of experience as an attorney/engineer in consulting, private industry, and government. He has personally designed, managed, and conducted more than 400 compliance, acquisition, divestiture, and specialty performance audits for a variety of industries, including several multiple site, fast track assessments for multimillion dollar transactions. He has also evaluated corporate EHS management systems and assisted in implementing improvements. Mr. Pooler brings an integrated legal, technical, and operational/managerial perspective to client assignments, and specializes in developing strategic approaches to environmental, health and safety management issues.

## Richard B. Storey

Richard B. Storey is a partner in the metropolitan New York business consulting practice of Arthur Andersen with more than 13 years experience. For the past three years he has worked exclusively directing activity-based management engagements, including work simplification, process reengineering and activity-based costing studies. He is also currently leading the firm-wide development of a performance information package for the telecommunications industry incorporating activity-based information as a key basis for measurement.

## Thomas F. P. Sullivan

Academically, Thomas Sullivan has a B.A. in philosophy, a B.S. in physics and math, and a J.D. in law. He has been in the forefront of the environmental field since the 1960s. He gained experience in industry before practicing law and representing clients in the environmental field. In 1973 he founded Government Institutes, and has authored and edited more than 100 books including *Environmental Law Handbook, The Greening of American Business, Environmental Health and Safety Manager's Handbook,* and *Directory of Environmental Information Sources.* He is a regular lecturer internationally on environmental topics and serves as president of Government Institutes.

## Gary B. Unruh

Gary B. Unruh consults on strategic safety, occupational health, and environmental issues for Conoco Inc., and DuPont businesses. Mr. Unruh has worked in the information management field for fourteen years. For the past four years, he has specialized in safety, health and environmental information management initiatives across the corporation. He received his B.S. in data processing from Emporia State University.

## Jack Weaver

Jack Weaver, Ph.D., is director of the Center for Waste Reduction Technologies (CWRT), a division of the American Institute of Chemical Engineers (AIChE). He also directs five other technical centers within the AIChE, including the Center for Chemical Process Safety (CCPS) and the Design Institute for Physical Properties Data (DIPPR). Prior to joining CWRT, he was a business manager and chemical engineer with more than twenty-five years of corporate business and technical management experience within the specialty chemicals, plastics, and petroleum industries, including international corporate leadership of safety, health, engineering, product safety, and environmental affairs for the Rohm and Haas Company. Additional experience includes leading a corporate pollution prevention program, developing corporate policies and procedures, waste minimization, life cycle costing, process safety management, business risk management, and decision-making. As an executive management consultant for Roy F. Weston, Inc., he assisted corporate clients in establishing and assessing overall management systems for safety, health and environmental affairs. He has also assisted clients in pollution prevention. Dr. Weaver holds three degrees in chemical engineering from Cornell University and the University of Delaware, and has also conducted postgraduate research in physical chemistry at the University of Muenster, Germany.

**Gordon A. West**

Gordon A. West is a senior consultant on environmental management for Pilko & Associates, Inc. At Pilko, his practice includes strategic and organization planning, risk management and benchmarking. His background includes extensive industrial management experience as well as environmental consulting. Prior to joining Pilko, he served as advisor to Arthur Andersen & Company and directed the environmental management consulting practice for Roy F. Weston, Inc. His work has included a wide range of safety, health, and environment management assignments in both the public and private sectors. His industrial experience was gained with the Rohm and Haas Company and with Corning Glass where he held positions in marketing, manufacturing, and senior staff management. At Rohm and Haas Company, he organized and directed their Regulatory Affairs Department with international responsibility for environmental control and product stewardship. Mr. West received his masters in business administration from the Wharton School of the University of Pennsylvania and a bachelor of arts degree from Trinity College. He has served on a number of industry and government boards and committees.

## Ray E. Witter

Ray E. Witter is a chemical engineer with more than 37 years experience in engineering, construction and manufacturing. Since his retirement from Monsanto Company in 1986, where he served as the company's director of safety & property, Mr. Witter has served as staff consultant to the Center for Chemical Process Safety of the American Institute of Chemical Engineers, staff administrator for the CCPS texts on *Guidelines for Technical Management of Process Safety, Hazard Evaluation Procedures, Auditing Process Safety Management Systems, and Documentation of PSM*. He is a lecturer for continuing education of AIChE for process safety management and auditing PSM seminars.

# 1

# Understanding The Company

*Gordon A. West, Pilko & Associates, Inc.*

## INTRODUCTION

The design of an Environmental Health and Safety (EH&S) program begins with an understanding of the company and the businesses that the program is to serve. The degree of sophistication and the level of effort required vary considerably among companies. The dimensions of the program are determined first by the potential impact of the firm's operations and technologies on health and the environment and how this potential has influenced the values and vision that have been established for the company; and second, by the size and complexity of the organization. Companies whose manufacturing and marketing operations involve the production and use of hazardous materials and who have concluded that a worthy and profitable future is dependent upon prudent and preventive management, will invest more heavily in specialized staff and control technologies. Companies with diverse business portfolios and global operations will deploy responsibility and resources more broadly. The operational and organizational factors that affect the design of an EH&S program are discussed in more detail in the balance of this chapter.

1

## OPERATIONS

The characteristics of a company's operations that are most important to the design of an EH&S program are

- market sensitivity;
- stakeholder sensitivity;
- material risks and impacts;
- history of problems.

## Market Sensitivity

Securing a major market share is an objective of most businesses. If these markets, and the production facilities required to serve them, are sensitive to environmental and health issues, business strategies will require that a higher level of attention to EH&S issues be given.

How well a company's products perform in comparison with competitors', in terms of environmental and human health impact, can have an immediate as well as long-term effect on market shares. Uninhibited by due process constraints, the marketplace can and does move faster than regulatory agencies in limiting the sale of products. If the environmental or health effects of a purchased product adversely affect the manufacturing or marketing of a customer product, purchases of that product will terminate as soon as a more satisfactory alternative is identified. Supermarket chains can halt—and have done so—purchases of specific food products because of newly discovered ill effects of pesticide residues. Their decisions have resulted in immediate termination of sales at the retail, distributor, food processor, and farmer levels of the market.

All customer decisions, of course, to halt product purchases or to change suppliers are not made instantaneously. An increasing number of companies, however, have active programs aimed at reducing use of and dependency on products and materials which pose potential health and environmental problems. These programs not only consist of looking for existing products that can be substituted, but also include development projects to solve the problem through new technology. Suppliers who are alert to such changing requirements and who have the resources to support their customers in development work have an important competitive advantage.

How well a company's facilities perform in satisfying current standards of environmental management performance also can affect market shares. If a supplier's environmental control practices do not satisfy a customer's total quality management (TQM) requirements, that supplier may lose out to a competitor whose performance does meet those standards. Environmental quality management (EQM) is becoming an integral part of TQM. The wide spread adoption of the International Standards Organization (ISO) Standard 9000, establishing criteria for TQM and securing certification for having met them, is about to be supplemented by a new EQM Standard - ISO 14000. Companies around the world are expending considerable effort and money to acquire ISO 9000 certifications for their facilities because certification is seen as positively affectively their position in world markets. Companies are expected to view ISO 14000 certification the same way.

## Stakeholder Sensitivity

Recognition of the environmental and health management expectations of company stakeholders has lead to well-articulated commitments to meet them. Companies making such commitments, however, are obliged to follow through on them successfully and consistently. Failure to appreciate the high level of concern about the environment that is held by people around the world (including investors customers, employees, and local communities) can lead to sorrowful consequences to the value of a share of stock, the marketability of products, the ability to attract and maintain a high caliber work force and to retain acceptance of operations by local communities. Proclamations of virtuous policies and practices, however, without first having in place the programs and management processes required to assure that they can be carried out, can magnify the problem at the bottom line. A well-designed public relations program results in elevated expectations and, without the supporting systems in place and demonstrated performance history, provides cannon fodder to the critics and skeptics anxious to blast management credibility and the corporate image.

## Material Risks and Impacts

The extent to which a company is dependent upon the use of hazardous materials in production and marketing is an obvious, major determinant in the design and management of an EH&S program. Producers of such materials have to commit significant resources to their programs. But purchasers also have to be concerned about how these materials are handled, sold, and disposed. The burden that falls upon all is to have the ability to assess and understand the risks associated with the use of, and exposure to these materials; to understand and apply the technologies, standards and, regulations that control the use and potential impacts of the materials; and to seek and find ways to eliminate or at least minimize use and impact.

## Environment Health and Safety Operating and Capital Costs

EH&S program costs can be enormous—a fact that is hard to accept in an era of intense global competition, re-engineering and downsizing. These costs vary between industries and among individual companies within those industries. Some examples from published reports for the year 1993 follow.

None of these companies, incidentally, project reductions in either capital or operating expense in the near future.

- Company: Rohm and Haas
- Industry: Chemicals
             ($ millions)
    Sales; 3,269
    Earnings; 126
    Total Capital Exp.; 382
    Environ. Oper. & Maint. Exp.; 105
    Environ. Equip. Cap. Exp.; 55
    Waste Site Remed. Accrual; 57

 - Company: Champion International
 - Industry: Paper and Wood Products
             ($ millions)
    Sales; 5,069
    Earnings; (156)
    Total Capital Exp.; 491
    Environ. Oper. & Maint.  Exp.; NA
    Environ, Equip. Capital Exp.; 71
    Waste Site Remed. Exp.; 5

- Company: Sun
  Industry: Oil
             ($ millions)
    Sales; 9,180
    Earnings; 288
    Total Capital Exp.; 612

Environ. Oper. and Maint. Exp.; 108
Environ. Equip. Capital Exp.; 123
Waste Site Remed. Exp.; 53

- Company: Bristol-Myers Squibb
  Industry: Health Care
  　　　　　($ millions)
  Sales; 11,413
  Earnings; 1,595
  Total Capital Exp.; 580
  Environ. Oper. and Maint. Exp., NA
  Environ. Capital Exp. including Waste
  Site Remediation; 50

The greater the impact of operating expense on earnings and the greater the diversion of capital away from productive uses, the greater the need to have a strong EH&S program, to eliminate materials and processes that require such expenditures and to establish performance measures to monitor ongoing cost-effectiveness. All four of these companies have strong EH&S programs.

## History of Problems

A number of the strongest EH&S programs came into being because of an event or situation which was both expensive and embarrassing to a company. The commitment to sustain these programs is being made, in part, to limit the possibility of the occurrence of another or similar event. These companies know what the impacts can be on the bottom line, market position, and absorption of senior management time and attention.

A number of the "events" that occurred, particularly in the decade of the '70s, drew wide publicity as part of the political drumbeat to gain enactment of environmental, health and safety legislation. It was in the '70s that senior managers in industry learned that public interest in environmental control was both keen and angry. They also learned that clear direction and motivation regarding EH&S standards and expectations had to be provided down through line management. Shortcuts and imprudent cost-cutting decisions made in small corners of a global corporation can have awesome consequences for the corporation as a whole. A company that has had a serious problem in the past and still uses materials and processes that pose potentially high risks will devote a high level of resources to the EH&S program and to internal communications.

## ORGANIZATION

Organizational factors that affect the design of an EH&S program are

- number and diversity of businesses;
- global operations;
- business unit autonomy;
- number and diversity of facilities;
- acquisition program.

### Number and Diversity of Businesses

A company that is in one business (as defined by technology, market, and strategy) will concentrate EH&S program resources and direction at the corporate level. "One size fits all" policies and programs can be designed to meet the needs of facilities: coordination with business strategies is feasible; technology changes can be followed and managed centrally; information management needs are fairly homogeneous throughout the enterprise; communication channels between the corporate headquarters and individual operating facilities are relatively uniform and stable; and government policies and actions can be monitored with a good understanding of their effects on companywide operations.

A company that is in a number of businesses with differing technologies, markets and strategies faces a much more complex problem in establishing an EH&S program that meets corporate standards while satisfying the needs of individual businesses. A company's business portfolio could include businesses of widely varying size, staff resources, technologies and rates of technology change, usage of hazardous materials, EH&S capital requirements, and configuration and location of facilities. In such circumstances, a centrally controlled and resourced EH&S program is difficult to carry out. While corporate direction is required to establish policies and standards and to make sure that they are satisfied, some deployment of EH&S resources is required. If program goals are to prevent problems from occurring and to control costs, the program must be integrated with, and responsive to the strategic and technological needs of the business. The trick is to deploy the EH&S staff in such a way that it is responsive, but not "Balkanized" or redundant.

## Global Operations

Until recent years, the environmental control "problem" has been regarded as primarily a United States problem. Certainly, the U.S. "Superfund" law stands out around the world as the most horrendously expensive legal quagmire that has resulted from the environmental movement. In addition, the propensity of U.S. society to engage in litigation has been encouraged by statutory rights for third parties to bring actions against alleged violators of environmental laws. Opportunities for such private actions are easy to identify in the United States because of federal, state, and local laws requiring public disclosure of environmental and health data and activities. As a consequence, many EH&S programs have had a United States focus.

The environmental movement, however, is a global movement, just as most industrial markets are global markets. Companies that are moving products, services, and capital around the world must now understand and comply with myriad regional, national, and local laws and standards. Evolving EH&S programs and practices of multinational companies tend to be fairly uniform globally, reflecting company values and standards. Nevertheless, there are differences in legal and marketplace demands that must be understood and addressed. In addition, there are initiatives being taken, particularly in Europe, that are destined to become international requirements for doing business. The ISO Standard 9000, governing quality management certification, and now the ISO 14000, proposed standard for environmental management registration, are examples of European initiatives that can affect a company's position in the world marketplace.

## Business Unit Autonomy

The concept of the strategic business unit (SBU) has been adopted to meet the need for closer integration of the functions involved in establishing and implementing the strategy of a business. The smaller structure, with a high degree of self-sufficiency and autonomy, is regarded as being "leaner and meaner" and more accountable for maintaining and expanding market share and for meeting corporate performance expectations. The extent to which the SBU form has been adopted by a company, or the extent to which its adoption is planned, affects significantly the design of the EH&S Program.

EH&S program managers and their staff are a support group. They advise line managers regarding EH&S requirements. If the advisors are in one place in the organization and people making business decisions affecting the environmental impact of operations are in another, the advice may not be given or heard in a timely manner. Managers accustomed to making decisions without consultation with corporate headquarters may not be disposed to accept advice from EH&S personnel who are not "on the team." Environmental management must become a close part of business management to avoid losing opportunities to be proactive and preventive and to minimize the chance that an independent decision by an SBU affects adversely the reputation of the entire company.

## Number and Diversity of Facilities

The manufacturing facility is the principal location where activities are taking place that can impact the environment and human health. This is where accidents and incidents are most likely to occur; where personnel training is most needed; where recordkeeping and reporting systems are needed to satisfy government requirements; and where pollution prevention and process safety must be addressed. This is also where line managers and EH&S specialists have the least time and freedom to be creative and plan ahead; to experiment with alternative materials and processes; or to read the *Federal Register*, state bulletins and those of foreign governments to learn about the regulations with which they are supposed to be in compliance.

Manufacturing facilities need legal, technological, and programmatic support at the same time that they need to be as self-sufficient as possible. Legal requirements need to be translated into terms, or standards, that are relevant to the operations of the facility. Alternative solutions to new requirements need to be developed and presented with sufficient lead time that they can be tried out and adopted.

The greater the number of facilities and the greater their diversity in terms of materials, processes, and location in the world, the greater the program challenge. The larger the number of facilities, the larger the potential for human errors and accidents. The greater the diversity among them, the greater the difficulty in the corporate development of control standards, pollution prevention and process safety approaches that are timely and practical.

## Acquisition Program

The assumption today, when acquiring an industrial company or business that has been in operation for more than ten years, is that the entity being acquired has environmental problems. The obvious question to be answered through due diligence is, "How bad are they?" How bad they are could determine the wisdom of making the acquisition or could reduce the amount of the purchase offer. The less obvious, but perhaps more important question is, "How well are these problems being managed?" In all but rare cases, the decision to go forward or not to go forward with an acquisition deal is made because the deal makes good or bad business sense to both parties, not because of the presence or nonpresence of EH&S problems. If the deal does go forward, the EH&S manager is presented with the task of handling a new set of problems through a new set of people.

The company or business being acquired may or may not have the same values and standards as the acquiring company. It may or may not have programs and systems in place that are similar to, or compatible with those of the acquirer. It may have technologies and serve markets that are unfamiliar to the acquiring company's EH&S staff.

The importance of implementing EH&S programs in a way that is sensitive to the needs of individual businesses is magnified in the case of a newly acquired business. Companies with active acquisition strategies need to involve EH&S staff early in the process, first to assess the nature and magnitude of environmental, health and safety problems and second to develop the plan and approach to bringing the new entity into the corporate program.

## SUMMARY

An understanding of the unique characteristics of a company—the environmental, health and safety risks and sensitivities of its operations and its organizational structure—is essential to the design of an EH&S program that interfaces with line management at the right levels with the right resources. The chapters that follow, address the most important issues and the approaches to dealing with these issues that face Environmental, Health and Safety managers and professionals and the businesses they serve. The importance of each of these issues to every company as well as the best way to implement the approaches suggested will vary from company to company. The value, however, of the experience of others cannot be understated.

The commentaries presented in the following pages reflect the experiences of companies whose EH&S programs have required a significant investment over a long period of time. An understanding of what these companies are doing and why they are doing it will be helpful in the design of EH&S programs for companies, including those with lesser needs.

An assessment of the level of EH&S effort that is required and the best way to deploy EH&S resources within the company can be aided by an examination of companies through a benchmark study as discussed in Chapter 15, whose operational and organizational profiles place them in the same quadrant of the EH&S matrix depicted in Figure 1. This matrix summarizes the operational and organizational factors that affect EH&S program design for all companies.

## Figure 1.
### Environmental Health And Safety Program Matrix

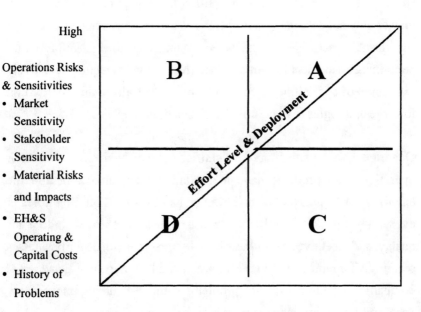

High

Operations Risks & Sensitivities
- Market Sensitivity
- Stakeholder Sensitivity
- Material Risks and Impacts
- EH&S Operating & Capital Costs
- History of Problems

Low          High

**ORGANIZATION COMPLEXITY**
- Number & Diversity of Businesses
- Global Operations
- Business Unit Autonomy
- Number & Diversity of Facilities
- Acquisition Program

- Quadrant A companies have diversified, global businesses whose operations and products have significant impact on the environment. How well they manage their EH&S programs can affect shareholder value. Their programs are expensive and their EH&S staff are deployed widely to support their businesses.

- Quadrant B companies are not diversified and are not international. They may or may not be large. Their operations and products, however, have a significant impact on health and the environment and their EH&S Program performance can affect the bottom line. Their programs require a relatively high level of commitment, but can be more centralized and focused on fewer technologies than companies in Quadrant A.

- Quadrant C companies have diversified, global businesses, but their operations and products have a minimal impact on health and the environment. EH&S problems, however, have a way of burning holes in everybody's pockets. Problems such as asbestos removal, indoor air quality, and repetitive stress illness have become major ones for businesses seemingly far removed from the environmental storm center. Quadrant C companies need to avoid such problems without incurring unnecessary costs and burdening management with irrelevant concerns.

- Quadrant D companies have small organizational challenges and small impact on the environment. This book will have greater value for those less fortunate.

The appropriate level and deployment of resources to achieve the objective depends upon the position of the company on the EH&S matrix.

The programs and strategies for EH&S management that are discussed in the following chapters represent the practices of companies that have well established EH&S management programs. These programs and strategies are designed to reduce environmental and health costs and risks and thereby to enhance shareholder value. This is a valid objective for all companies in business today.

# 2

---

# PROGRAM DRIVERS

*George J. Miller, Dechert Price and Rhoads*

## WHY DO ALL THIS?

A principal question raised by corporate executives faced with the challenge of cost and staff cutting in redesigning the corporation is whether or not the cost of an environmental management program is money well spent as compared to the many other competing candidates for corporate expenditures. The answer lies in the many long-term advantages such a program will bring to the company, its stockholders, and executives. These advantages include

1. Long-term asset preservation;
2. Personal and corporate liability protection;
3. Product marketing advantages;
4. Good community relations;
5. Operating and maintenance cost savings.

## ENFORCEMENT HISTORY

EPA enforcement statistics serve as a barometer of the nature of the problem. In fiscal year 1994, EPA collected a record $164 million in civil and criminal penalties—up from $145 million in fiscal ear 1993 and 120 million in fiscal year 1992. An additional $1.4 billion was pledged by responsible parties at Superfund sites. Of more personal interest to executives, convicted polluters were sentenced to a record 99 years in prison.

The cost of responding to EPA's enforcement efforts is extremely high. EPA brought a record 2,247 enforcement actions in fiscal year 1994 that resulted in sanctions, an increase of 28 percent from 1991. Of these 1994 enforcement actions, 220 were criminal actions, 1,597 involved administrative penalties, 403 were civil cases referred to the Department of Justice, and 27 cases resulted in consent decrees.

Unfortunately, there are no overall statistics on enforcement by state regulatory environmental agencies or on the virtual flood of private cost recovery actions brought by disappointed purchasers of corporate assets under the federal Superfund law. More important, there are no statistics on the value loss of corporate assets as a result of hazardous materials mishandling in the manufacturing process. Frequently, the most important consideration in a corporate acquisition is how to deal with environmental liabilities that can decrease the value of otherwise valuable business properties.

## LONG-TERM ASSET PRESERVATION

### Destructive Liability

The critical asset of any manufacturing company is its manufacturing facilities. Sloppy practices in the disposal of waste or in permitting spills in the conduct of its operations may lead to extremely high remediation costs over many years with a resulting diminishment of value in those facilities. Under the federal Superfund program (the Comprehensive Environmental Response, Compensation and Liability Act or "CERCLA)," the owner or operator of a manufacturing facility is jointly and severally liable for the costs to cleanup areas contaminated by hazardous materials at the manufacturing facility or other places where the contamination from the facility may have been placed or to which the contamination has migrated. This liability also extends to any person who has arranged for the disposal of hazardous waste. This liability is retroactive to activities prior to the 1980 enactment of Superfund and may be asserted by EPA, state governments and private parties. Since the responsibility is imposed on the owner of the facility, the likely cost of meeting cleanup responsibilities is a major element of any negotiation on the sale of those facilities.

This program has been a nightmare of legal and engineering costs with comparatively little being spent on successful remediation. This has been caused in part by the extremely strict cleanup standards that EPA has required for remediation activities. In addition, remediation has proved to be much more difficult and expensive than anyone had realized. Once hazardous contaminants have escaped into the ground water, meeting remediation goals requires an enormous capital expense for remediation facilities and a nearly interminable obligation for operating and maintenance expense. The cost of operating and maintenance expenses for ground water remediation is generally estimated over a 30-year period based on cost of money considerations, not because it is expected that the ground water will meet cleanup standards by that time. Liabilities for cleanup may well exceed $80,000,000, without considering the extended period of operating and maintenance expense of the remediation facilities.

Each state typically has either a hazardous waste cleanup act of its own, under which cleanup of hazardous waste sites may be ordered. In addition, each state typically has other laws aimed at protecting soils and ground water from contamination from industrial wastes, which, in many cases, extend beyond what may be categorized as hazardous materials under the federal Superfund program.

If the government doesn't catch up with you, private citizens may. As noted above, the Superfund liability may be asserted by private citizens and many of the federal and state environmental laws permit private parties, including environmental organizations, to bring actions for cost recovery or penalties for violations of the applicable environmental law. Recoverable penalties for violation of a water discharge permit alone have exceeded $4,000,000. Environmental groups which bring these actions are typically inflexible in approaching settlement of their claims for penalties and recoverable counsel fees and litigation costs. Accordingly, the costs of litigation against these groups may be very high.

## Asset Sales

Superfund and state environmental liabilities have an adverse effect on sales and purchases of assets. Frequently a choice must be made between accepting a much lower purchase price and giving an indemnity for uncertain future environmental responsibilities, with a fund deducted from the purchase price as security for the indemnity. The cost of required environmental assessments of dirty properties will be high. Further, a prospective purchaser's investigation of the property may disclose contamination that may have to be reported to government agencies with the consequence of the imposition of cleanup responsibility. Even if the transaction is consummated, there is always exposure to suits from the purchaser for improper disclosure of environmental liabilities in connection with the transaction.

## Protection of Stock Value

All public companies must meet SEC disclosure requirements with respect to environmental liabilities. These disclosure requirements are becoming increasingly complex. This only increases legal and accounting costs, to say nothing of the expenditure of management time. Associated accounting principles may require the creation of reserves to meet these liabilities.

The failure to deal properly with disclosure of these liabilities may lead to exposure from the securities' class action artist. A significant number of the plaintiffs' securities bar have targeted disclosure of environmental liabilities as a new large market for their services.

Stock value will be adversely affected by a disaster of a high visibility liability situation. Releases of hazardous materials which result in public health problems results not only in significant liability but is also likely to result in a decrease in the company's stock value. The experience of the Bophal disaster is only an extreme example. A proper compliance program may avoid any such disaster and will be designed to comply with emerging requirements under the Clean Air Act and state implementation of disaster prevention plan requirements through air permit requirements.

## Facility Location and Purchase

The development of an environmental compliance program also has the benefit of keeping the company alert to environmental requirements which may have a marked effect on the value of the company's assets. The following are only a few examples:

1. Decision on the purchase of new manufacturing equipment should be informed of available options to reduce the creation of pollutants in the manufacturing processes;

2. Decisions on the purchase of pollution control technology should be informed of requirements for the selection of the best available technology in order to avoid, as much as possible, retrofit requirements.

3. Decisions on the selection of facility locations should be informed of environmental requirements. Geographical areas which are classified under the Clean Air Act as "moderate," "serious," "extreme" probably should be avoided because regulatory costs of permitting and new facility siting are high; wetlands and floodplains present extreme regulatory difficulties, and location near highly protected waters or an inadequate sewage treatment plant may well increase the cost of regulatory requirements and the cost of waste water disposal.

## PERSONAL AND CORPORATE LIABILITY PROTECTION

A second major reason for having an adequate environmental management program is to protect the company and its managers from criminal or civil liability. The environmental liabilities discussed above may be imposed not only on the company, but also on those managers having responsibility for the management of the company's activities that result in violation of environmental requirements. While this is particularly true of those individuals who are actively involved in the disposal of waste from the company's manufacturing operations, liability may also be imposed on managers who themselves have acted properly but have been in charge of company operations which have resulted in environmental violations.

## Responsible Corporate Officer Doctrine

Perhaps the most dangerous legal principle in this connection is the so-called "responsible corporate officer doctrine," under which liability may be imposed on a corporate officer or manager only because the violation occurred on his watch, not because he did anything that was wrong. Both the federal Clean Air Act and Clean Water Act provide for criminal responsibility for "responsible corporate officers," although the statutes do not define this term. The Clean Air Act provides for civil and criminal enforcement against owners and operators and defines "operator" for this purpose as "senior management personnel or a corporate officer." Both Pennsylvania and New Jersey statutes provide for liability for responsible corporate officers. Even without these explicit statutory provisions, court decisions in the environmental field over the past 15 years have applied this principle in finding individual liability for violation of environmental requirements.

An officer's studied ignorance of activities by others in his supervisory control is not likely to be a defense. For example, the criminal provisions of the Clean Air Act permit the inference that the officer knew of the violation through proof that he "took affirmative steps to be shielded from relevant information." The United States Department of Justice will say that this is merely a restatement of existing law applicable to nearly all criminal cases.

## Certifications of Compliance

Perhaps the second most dangerous exposure to criminal liability is in the increasing requirement that officers or managing agents provide certifications of compliance for permit applications and, from time to time, for ongoing operations. The application for a major source permit under the Clean Air Act requires an officer or managing agent to certify that the company is in compliance with "all applicable requirements" of the Clean Air Act and implementing regulatory programs. Just understanding what those requirements are is beyond many lawyers and environmental managers, to say nothing of whether the company is in compliance with those requirements.

The person called upon to sign such a certification has jokingly been referred to as "the designated felon" because of his exposure to criminal enforcement for making a false certification. A similar certification is required by the permitting regulations under the Clean Water Act and the Resource Conservation and Recovery Act. In addition, EPA is demanding provisions in Superfund consent decrees for similar certifications with respect to the remedial investigation reports and feasibility studies submitted by a responsible party as part of its obligation to remediate a contaminated site under the Superfund program.

The liability exposure for false statements can be substantially reduced if the company has an effective environmental compliance program on which the officer or managing agent can rely in making the required certification. In today's environment, in which prosecutors believe that an effective environmental compliance program is clearly standard practice, it will be difficult to defend false certifications where the company has not made an effort to establish such a program. In addition, a bad environmental record may only make such a prosecution an easier task.

## Federal Sentencing Guidelines

The problems of criminal enforcement are aggravated by the felonization of most environmental crimes, at least at the federal level. All but one of the criminal violations under the Clean Air Act Amendments of 1990 are felonies. Accordingly, the time-honored practice of making a deal with the prosecution by agreeing to plea to a misdemeanor, has become increasingly difficult in connection with environmental violations. Obviously, a felony carries a higher sentence and entering a plea to a felony may have very serious consequences under the federal sentencing guidelines. It will also act as a bar to the company's contracting with the federal government in its procurement programs—a possible loss of major revenue for the company.

The federal sentencing guidelines arising out of the 1984 Sentencing Reform Act may soon be applicable to environmental crimes by organizations. The object of these guidelines is to restrict the discretion of federal judges in sentencing by using a formula to ensure that the sentencing for federal crimes is comparatively uniform throughout the country. Under the recommendations of the Sentencing Commission's Advisory Group on environmental crimes released in March, 1993, base fines are proposed which are the greater of (1) the economic gain to the organization plus costs directly attributable to the offense, or (2) a percentage of the maximum statutory fine that could be imposed for the offenses based on a table providing different percentages for different types of offenses which vary with the seriousness of the offense. The base fine is subject to various aggravating and mitigating factors which may increase or decrease the base fine by specified percentages. Aggravating factors include

- Management involvement in the offense;
- The organization's criminal compliance history;
- The organization's prior civil compliance history;
- Any concealment of the offense;
- The violation of any existing order; and
- The absence of a compliance program or the failure to implement an existing program.

Specific mitigating factors include

- A demonstrated commitment to achieving and maintaining compliance with environmental requirements;
- Compliance and self-reporting; and
- Prompt remedial assistance to victims.

## Environmental Management Programs

Not surprisingly, these proposed sentencing guidelines focus heavily on the existence, structure, and implementation of environmental compliance programs by organizations. The elements of such a program are described in detail in Chapter 4.

Similar guidelines will apply to individuals. The advantages to the corporate officer charged with a violation—despite the operation of an effective environmental management program—should be obvious. An officer may or may not be protected by a provision in the company's by-laws that indemnifies the officer from litigation. However, most state corporation laws will not permit indemnification for willful violations or criminal acts. Accordingly, if the officer acts without the benefit of an adequate compliance program, he may find he will incur both defense costs and applicable fines.

In the case of civil liability, the failure to keep adequate records can alone be the basis for civil liability. Both the environmental and OSHA laws and regulations contain many recordkeeping and reporting requirements which must be strictly observed. Nearly all civil enforcement of the environmental laws involve claims that appropriate recordkeeping requirements and reporting procedures have not been followed. This includes recordkeeping requirements with respect to employee training.

The implementation of a proper environmental compliance program will not only ensure the right action is taken, but also that it is properly recorded for purposes of any enforcement action. Records as to when environmental infractions cease are also important. Under the Clean Air Act, for example, a violation is presumed to continue until the defendant is able to establish that the violation has ceased. Since penalties are frequently based on a dollar amount per day of violation, having a record as to when any violation ceased is important.

Voluntary disclosures of violations may also be a benefit of an environmental management program. Under EPA's recently announced interim policy for voluntary environmental self-policing and self-disclosure, companies with voluntary audit programs can eliminate or reduce the potential for certain civil penalties for voluntarily disclosed violations. This policy was announced on April 3, 1995 and may be found at 60 FR 16875-79. While the EPA policy rejects the concept of a self-evaluative privilege for environmental audits, a significant number of states have enacted statutes creating an "audit privilege" designed to encourage monitoring, discovery and disclosure of violations by preventing the results of audit programs from being used in various contexts, such as citizen suits and toxic tort claims.

Exposure to personal or corporate liability is not limited to violations of the environmental laws. As indicated above, environmental liabilities have implications under the securities laws. A failure to make proper disclosure of environmental liabilities may not only lead to difficulties in required financing for expansion, but may also lead to possible personal liability in litigation brought for failure to make proper environmental disclosures. The securities class action bar is well aware of this potential for very profitable litigation. A proper compliance program should assure compliance with disclosure requirements.

## PRODUCT MARKETING ADVANTAGES

A most important advantage of an adequate management program is that it enables corporate executives to recognize opportunities for product development and product marketing. While environmental regulations are most frequently viewed as a drag on profits, environmental laws sometimes provide product development and marketing opportunities which otherwise might not be recognized. Perhaps the most well known example is the development of the catalytic converter for automobiles, which was required in the early days of the Clean Air Act in the 1970s. While the Clean Air Act's ban on the production of chlorofluorocarbons is bad news for those dependent on their sale, EPA's program for identifying useable substitutes for these products is likely to be good news for others. Similarly, EPA is developing a program to regulate products containing volatile organic compounds (VOCS) such as paints and architectural coatings which may remove many products from the market. On the other hand, those who have recognized the market for water-based paints or water based metal coatings as substitute products may profit immensely.

The need for air pollution technology and better combustion technology to reduce emissions of VOCs and nitrous oxides has been well recognized and exploited by many companies. Recent regulatory programs aimed at reducing pollution at the source will promote the development of more efficient industrial processes, which may provide a fertile ground for the development of new products.

## Green Advertising

There are also many opportunities for "green advertising" that can be exploited with the proper development of an adequate environmental management program. Much green advertising has been found improper by the FTC or by the courts because the advertised claims have not been properly based in fact. Sensitivity to environmental compliance will not only promote proper green advertising claims, but will enable a company to obtain an environmental seal of approval for its products based on its record of environmental compliance.

EPA's leadership program, which would permit such an approach to advertising a company's compliance with environmental laws in connection with its products, is still in very formative stages. However, regulatory developments in Europe are moving into an advanced stage.

## ECO-Label

The EU Commission's proposal for an EU ECO-Label award program, under which companies may advertise their compliance with environmental requirements in connection with marketing their products, was adopted in June, 1992. This program applies to all products except food, drink and pharmaceutical. Competent bodies in most member states have been established to carry out this program. Product criteria for the first several product groups (such as detergents, paints, varnishes, washing machines, dishwashers, shellacs, batteries, light bulbs, construction materials, fertilizers, insulating materials, toilet paper, kitchen towels, absorbent paper and offset printing paper) are being established so that applications for award of the Eco-Label for products in these groups may be adopted by member states. In this connection, environmental management auditing standards are being developed by the European Committee for Standardization (CEN).

National environmental management standards exist in the UK, Ireland, Spain and France. They are similar because they are derived from a UK standard.

## ISO 14000

On an even broader base, the International Organization for Standardization (ISO) has developed international standards for environmental management, under the title of ISO 14000. This program may become extremely important to manufacturers of products or services which cross international borders. An existing ISO standard, ISO 9000, provides for certification of companies for making excellent products and backs them up with high quality service. It has been regarded by some as a valuable marketing tool.

ISO 14000 is designed to achieve several purposes which include making it more difficult for countries to use environmental issues as trade barriers, creating a universal set of standards to help businesses meet their commitment to the environment, and avoiding the need for multiple registrations, inspections and certifications as their products cross from country to country. ISO standards will address five major areas:

1. General corporate policies and procedures governing environmental management systems;
2. Environmental auditing;
3. Performance evaluation;
4. Environmental labeling; and
5. Life cycle assessment.

The guidelines on the environmental management and auditing portions of this program have reached "final draft" status and are expected to be formally adopted in 1996. (See Appendix at the end of this book.)

Compliance with these standards, certified by procedures similar to those of ISO 9000, may provide international marketing opportunities not open to those who fail to obtain a certification. U.S. industry participated heavily in the development of these standards. However, much of U.S. industry does not understand or have information on these developing standards. Organizations that want to stay competitive in the international market should focus on improving their environmental management programs before these standards are promulgated so that they are in a position to seek certification. If customers demand it, it is likely that some major companies will ask their suppliers to be ISO certified and nearly all companies will want to show proof that they are environmentally responsible.

The development of these environmental auditing standards will have enforcement implications. Liability frequently turns on whether a company has met the standards of the industry. While both the ECO label program in the EU and these developing ISO standards are meant to be voluntary, they are standards that may be relevant in both government enforcement actions and private party suits. A certification that the company has met these standards will definitely be helpful.

## GOOD COMMUNITY RELATIONS

Few EPA programs have had a greater impact on the willingness of industry to focus on pollution reduction than the Toxics Release Inventory (TRI) program which was created by the 1996 amendments to Superfund through The Emergency Planning and Community Right-to-Know Act. This act established elaborate provisions to assure that the community in which the manufacturing facilities are located will be informed of hazardous materials stored and used there and of spills of hazardous materials at these locations. Most important, this Act required companies to report annually on their generation and disposition of wastes.

The initial TRI report distributed to the public was a public relations disaster for the manufacturing industry because it portrayed industry as creating absolutely enormous amounts of hazardous wastes which made even major companies appear to be bad polluters. The public reaction to that report reached the highest levels of corporate management with the result that industry became interested in voluntary programs for the reduction of pollution such as the 33/50 program in which many companies pledged to reduce their generation of hazardous wastes by 33 and 50 percent by specified deadlines.

The principal driver of this program is the fear of accidental releases of hazardous materials as exemplified by the Bophal disaster and *Exxon Valdez* spill. Any incident of releases of this nature can have a marked, adverse effect on the company's reputation in the community and on the marketing of its products. In addition, good relations in the community are essential to obtaining local and state government approvals for many of the company's activities and plans for future development.

## OPERATING MAINTENANCE COST AND SAVINGS

The cost of a first-rate environmental management program is not to be underestimated. This may be particularly true in the initial years of the program, in which remedial action has to be taken promptly. These costs might be deferred into later years when the company's properties are sold, but many of them will have to be dealt with much sooner in connection with government enforcement programs or needed long-term financing.

Deferring dealing with most environmental management problems, however, will only make their solution more costly. A spill of hazardous materials today, cleaned up in a good management program, will be much less costly than a full ground water remediation program years from now, which may require 30 years more of expenditures to remedy.

Despite the feared initial startup costs, cost savings of an environmental management program will be of great financial benefit over time. Sensitivity to manufacturing processes which will reduce pollution at the source will reduce the heavy costs of waste disposal. Whether industrial wastes are incinerated or transported to authorized disposal sites, these costs may be unnecessary if enough thought is given to alternate production processes or raw materials. Further, compliance with environmental requirements will reduce legal expense, long-term remediation costs, and excessive penalties.

A most important consideration is the savings to be obtained by reducing executive time devoted to solving environmental problems. While training managers on the implementation of an environmental management program will have a significant initial cost, it will result in greater sensitivity to the benefits that can be achieved through meeting environmental requirements and taking advantage of them in devising new products and new manufacturing systems to reduce operating costs. Most important, once the management program has been established and is working effectively, less executive time will have to be spent on environmental matters.

# 3

---

# ESTABLISHING THE SCOPE OF ACCEPTANCE OF CORPORATE POLICIES AND PROGRAMS

*David R. Chittick, AT&T*

## INTRODUCTION

It should be clear from the previous chapters that there are several choices and consequences of company programs. The selection of a program and its creation is not as easy as it sounds for, as one person put it "...individuals are capable of assigning probabilities to alternative outcomes that are consistent with probability theory and contain all relevant information. However, there is a substantial body of evidence that people find it hard to think in probabilistic terms and often have serious misconceptions about the magnitude of important environmental risks."[1] Doing nothing is a choice, of course, but that is likely to be a very costly one. The object of this chapter is to examine the ways of establishing various programs and getting them into the framework of the business.

## QUALITY CONTROL PROGRAM TECHNIQUES

The underlying principle of Total Quality Management (TQM) is that all defects are waste. Thus, it follows that air and water discharges, as well as solid waste, are the result of process or system defects. What rational businessperson would procure material for manufacture and then throw part of it away? To be sure, this seems to oversimplify the problem of the environment, since there is inefficiency in every process, but there are great gains to be made by uncomplicated adjustments or changes to existing procedures. While not the subject of this chapter, the same can be applied to worker safety—for what greater defect can be found than that which causes an injury?

If a company has an established TQM program, the implementation of an environmental program can be easily incorporated into the existing structure. Before any program can start, however, there should be an understanding of what needs to be done. Chapter Two describes the reasons why a program is necessary and, in most cases, required. The old adage that half the solution to a problem is the definition is certainly true here. Typically, this is the hardest part of the total program, for several reasons. First is the complexity and sheer volume of educating management, at all levels, about the requirements that government places on business (government, in this sense, meaning that which is applicable to the location or locations and at all levels of government: local, state, provincial, national, etc.). Second is the cost of doing a corporate wide environmental survey or audit to determine the current status. In the competitive world, additional cost that does not add to the product or service is not looked upon with joy by management of any company. Third is the reluctance to getting a full understanding of the environmental status—the "sleeping dogs" or "if it ain't broke" syndrome. It is very common to hear people say they don't want to know about environmental problems, as if that somehow protects them. A variation of this theme—"known or should have known"—is put forth by company attorneys, and with good reason, (See Chapter 14, "The Role of Auditing"). Because of the litigious nature of U.S. law, all environmental reports, including audits, must be written very carefully and accurately by people who are trained to do so. Many employees are frightened by the unknown legal surroundings, which are unfamiliar to them. The management education process, therefore, is one that has to deal with the corporate culture, the geographical—and therefore legal—expanse of the corporation, and the situations which are known.

A first step is to educate management about the need and to gather an overall view of the company's environmental health. TQM principles can then be used to help identify the next steps. In this case, Pareto analysis is quite often applied to this data to help surface and rank the order of the largest problems. The 19th century economist, Vilfredo Pareto, pointed out that eighty percent of the problems come from twenty percent of the possible causes. The trick is to identify the top twenty percent. An overall audit of the entire corporation, which has been designed to give an accurate report of the current status will provide the base data. This audit phase requires a great deal of coordination within the company, because the environmental conditions are likely to vary between organizations and there will always be the question of who pays for corrective action. The object of this phase should be kept clear, however, and the audit data must be reduced to a well-thought-out rank listing of the company's overall environmental problems.

## GOAL SETTING

The ultimate goal for an environmental TQM program is total compliance with environmental regulations. This goal is somewhat akin to the TQM zero defect goal, for while "zero" is statistically unattainable, one can set programs in place to move in that direction. The same is true with total compliance with environmental regulations. For example, in the United States there are over 100,000 environmental regulations. Annually, at the federal level, there are in excess of 10,000 pages in the *Federal Register* which add, change, or modify them. It is difficult to envision a company that could have total compliance with all regulations at all locations around the world. "Indeed, two-thirds of the [top corporate] lawyers surveyed acknowledged that their businesses have operated, at least some of the time in the past year, in violation of federal or state environmental laws."[2] It is imperative that a company establish an environmental program to do just that—total compliance. By taking the rank order listing of problems identified by the Pareto analysis, it would be helpful to set solution-oriented goals with the following points in mind:

- Four goals are better than forty;
- What are the expected results?
- The goals support the company's long-term plans;
- The goals should be reaching.

"Four goals are better than forty" simply means that not all things can be done at one time. There are many corporate goals which require management's time and are often in conflict with one another. In establishing environmental goals (or any other goals for that matter), only a few high-level and well-defined goals should be proposed. Don't try to solve all the problems in one day because, quite simply, there is not enough money or management time to do it.

The "expected results" must be defined so that all employees have a clear understanding of what it means to the company and to their job. For example, if one of the major problems that comes out of the Pareto analysis is volatile organic compounds (VOC) emissions from painting operations, a goal such as "comply with the U.S. Clean Air Act" is not as meaningful as "reduce the emission of volatile organic compounds (VOC) by 95 percent by the end of 1999." For every goal, *there must be a metric by which progress can be measured*. But measurement for measurement's sake is costly and of limited use. What to measure and how the measurement relates to the goals and objectives of the business are extremely important. (For a thoughtful article on environmental measurement, see endnote [3]).

The goals must also fit into the company's long-term plans. The use of a VOC goal in the previous example is of no use if there are plans to eliminate that business from the company. The goal has to be real to the employees and something to which they can relate. They also must realize that nothing is static. Products and processes change and factories change as well. For example—again using the VOC goal—imagine two or three years into the program and one plant is closed, with the production shifted to another plant, so that the VOCs double with the increased load at the recipient factory. Corporate VOC use may decrease but the plant manager of the factory that received the work will see a large increase in VOC use and, therefore, the goal achievement becomes much more difficult to achieve. The long-term plans must also guess at future regulations and their impact on future processes. The United States is moving to a tolerance of fewer and fewer emissions of any kind and toward a public forum regarding the ability of a manufacturing facility to continue to discharge toxic or harmful material into the local environment.

Finally, the goals should be something which are a real achievement; a target which stretches the organization to reach and is therefore something to celebrate when achieved.

## MANAGEMENT'S ROLE

Every book and paper written on Total Quality Management states the absolute requirement of "top level commitment." Some CEOs are fearful that this means overextending themselves, since they already feel in overload. While top level support is critical, it doesn't necessarily mean that establishing an environmental program will place an unnecessary burden on the executives. In the United States, as well as many other countries, that burden is already there by statute and a good Total Quality Management program is an excellent way to carry out the requirements. There are many ways to establish a program but the one most likely to succeed is one that is designed by the company itself and fits its personality. In addition to goal setting, common to almost all programs are:

- Commitment;
- Policy Statement;
- Champion;
- Follow up.

*Commitment* means not only top level support in very visible terms but also in the commitment of money and resources to accomplish the tasks. As an example, there is an executive who scheduled fifteen minutes on the agenda to review the environmental goal progress at every location he visited. Since he was personally interested, it did not take long to generate interest in the rest of his organization. "Management's commitment must be carried throughout the corporation. The best way to emphasize this commitment is to state in the policy that the line managers are charged with individual responsibility for the environmental performance of their activities."[4]

It follows therefore, that the *corporate policy statement* is perhaps the most single important step. As Dennis Kinlaw stated, "The policy statement establishes the ground rules and expectations for the organization by:

- Clarifying for the organization exactly where the organization stands regarding the environment and its business interests;
- Integrating concerns for the environment with the strategic business interests of the organization;
- Focusing the organization's attention on those few interests that are crucial to its success."[5]

The statements of many companies may seem bland and dull, but to really serve a company well, they should be the product of a well-thought-out and deliberate program—one which includes the personal endorsement of the senior executives in the company. Not only must it say what the company principles are and what the company is going to do, but it must stand public scrutiny as if it were printed on the front page of the local paper, as well it might be. It becomes the focal point upon which all else is built.

Since every project needs a leader, a *champion* is needed to head the environmental goals project. It should be someone who can communicate with senior executives, can keep them informed of the progress or problems, and can push when needed. The total scope of the duties of the champion will become clearer when discussing selling the goals.

There are two additional points about goals. If a company is involved in global trade or has operations in more than one country, think how the goals (policy or anything else for that matter) will be interpreted and understood by others where English is not their first language. Many multi-national companies routinely send drafts to key contacts around the world in order to assure that the message is the same and to gain commitment. While the process is time-consuming at first, the result is a much better product that reaches achievement faster because everyone understands the objective and is committed to the result. (It also reinforces the fact that not all knowledge is at headquarters!)

The last point about goals is that the champion or organization at the company might have a title such as environmental control, but the goals *must* be adopted by those who have the ultimate responsibility. The line management can not delegate that part of the job outside the operational function. The next chapter will deal in depth with the subject of organization for environmental programs.

## FOLLOW-UP

### Selling the Goals—Internally

After the goals and policy have now been established and approved by senior management, no one will be surprised if nothing happens. Think of all those people in organizations throughout the company who have their own problems and are not looking for more. "The goal is total company involvement, from the board and the CEO level through every responsible employee."[6] The goals must therefore be carefully deployed throughout the company using the Total Quality Management process. This is where the champion comes in and becomes the chief salesperson. Indeed, the deployment of goals requires a very careful sales campaign. Target audiences and organizations should be identified and time schedules established. The messages must be clear and the goals stated in such a way that all affected groups can understand the necessity, adopt the goals, and act upon them.

It was previously stated that measurements were an absolute must for all goals since no quality program is possible unless measurement of progress can be made. In the deployment of goals within a company, the measurement criteria, techniques, and data collection system are also required. In the sales campaign therefore, the measurements should be fully explained so that all locations/organizations report the facts the same way. The annual air emissions from a location could be in tons per year, but does that mean 2,000 pounds per ton as used in the United States (also called the short ton), or 2,240 pounds per ton as used in some countries (called the long ton), or a metric ton, which most of the world uses? For a company with major facilities throughout the world, it may be useful to use the metric system for all measurements. Again, this is a corporate culture issue, but if a "directive" comes out of a headquarters in the U.S.A. to locations in Europe or Asia, loaded with requirements expressed in pounds and gallons, it's not likely to be received very well. The same is true of material loaded with United States government regulations and requirements.

Include in the measurement plans the fact that all reports are (or can be) public documents. Attempts to conceal or protect information will be seen as an attempt to hide something that is illegal. Perhaps the best measurements are those which are required by law or regulation. Since that data is required anyway, it poses little or no additional expense to the program. It is also public information, so those outside the company who might be interested can track the progress made against the established goals. Those naysayers who doubt the accuracy of the reported information should be reminded that, in the United States at least, it is a federal crime to knowingly provide the government with false information.

There is another incentive to use information that is required by regulation. Without an overall system approach, a company makes an investment in the collection of the required information. If all that happens to it is that a line on a government form is filled in, the investment is wasted. Only when the information becomes part of a database does it become useful to mark progress, to test the validity of the data, and to alert management to change. The data can also be combined and hierarchically oriented by organization as well.

The internal sales campaign is, then, the key to successful deployment of the goals. The measurement system becomes the incentive system to show progress and allows the champion to keep the linkages together. Successful projects are publicized and events are celebrated as a reminder and as a means of keeping the goals constantly on the minds of the employees. One example of a celebration at a factory was when a machine was changed from an organic solvent to an environmentally friendly water-based material. All the people in the area took part in placing a large sign on the machine that had a red circle and diagonal slash over large black letters on a white background that said "VOC." People took pride in the fact that "their" machine had made the goal. When the last machine in the plant was cut over to the water-based process and the entire factory was VOC free, the management hosted a plantwide celebration with coffee and cake for all.

As the goals are deployed, it is possible to chart the deployment by organization and location. Starting at the top, one should be able to go to any organization and work downward to the ultimate point of the goal—the cause or source of the environmental problem. At the root cause, there must be a detailed examination of the cause, the history of the events, the plans for correction, the time frame, the anticipated cost and savings, and the person responsible for the action. This information is valuable to understand the time frame and resources required to achieve the goal. If after gathering the information, it becomes apparent that the goal will not be achieved in the desired time frame, the champion is in a position to investigate and reallocate resources as needed. The important point here is that it is the local investigation, initiatives, and resources that are applied to solve the local problems. This contrasts with many corporate "solutions" which are promulgated from headquarters. Local solutions provide for innovation and cost effective changes which promote local pride. The ability to link corporate goals with local solutions is called by some TQM groups, "The Golden Thread." These threads weave the fabric of total corporate commitment. Once established, the tracking of progress and status becomes much easier and cost effective. By using the threads, the champion is able to push the deployment down each of the organization chains and to track problems and status.

With all the various problems facing the managers, it takes patience and salesmanship to keep the environmental goals alive. Like anything else, it requires constant care. It is important for the champion to make allies with various staff groups, the public relations group (to get space in the internal company newspaper), and any other connection that will work. This is a function of the corporate culture, of course, but the process is common—keep the goals in front of everyone. During the period of initial deployment, the champion can test the commitment by means of the budget and expense reports. This means a tie to the accounting organization, because when all is said and done, it is the commitment of funds for the people, the research, the equipment and all the changes that are required that make the difference.

## Selling the Goals—Externally

There is a growing demand throughout the world for corporate reports of environmental performance. The International Standards Organization (ISO) has been working for some time to establish a standard reporting technique. Other groups have, or are attempting to do the same. A company can use the annual progress against the goals as both an internal and external report. This external reporting process takes time to grow but is worth the effort. At some point in the development process, it would be worthwhile to discuss the annual report design and format with an environmental advocacy group. The exposure to different thoughts is very helpful and leads to a better product. Some environmental groups insist on including information such as fines, and ongoing environmental legal problems such as the number of "Superfund" sites, but this information tends to be very misleading since it is not a reflection of the current environmental activities of a company. It is the current and future activities that should be the focus of the program.

There are a number of external goals that can be used by companies in cooperation with governments. The U.S. EPA, for example, has programs designed to push results beyond the regulatory limits. These programs include the "33/50 Program," which pushes for a 33 percent and 50 percent reduction in the emissions of certain chemicals; the "Green Lights" program, which targets the replacement of energy-consuming illumination with new high-efficiency lighting; the computer STAR program; the U.S. Department of Labor's OSHA Voluntary Protection Program; and so forth. All of these programs take management time and expense, but there are returns for that effort and, in some cases, large returns.

For those managers who might think that there is no point in "going public" with the company's environmental status and that all it will bring is trouble and a lot of unfavorable publicity, that is somewhat true. The results of doing nothing, however, appear to be even worse because there will come the time in many parts of the world (for example, the USA and the EC) when there will be a public referendum to decide on air, water, and solid waste discharge permits, which will be required to continue operations at that location. It follows then, that the public must have trust in the environmental care of a company and this can only be done over time with frankness, openness, and honesty.

# SUMMARY

There is a great value to establishing environmental goals within the framework of a Total Quality Management process. A successful program lifts the total compliance effort, reduces risk and cost, and improves the reputation of a company. It is the reputation that takes so much time to build and which can be lost in one careless act. Zero defects don't just happen. They are earned! To paraphrase an old adage, "Without a set of goals, any road will take you there."

# ENDNOTES

[1]    Freeman, A. Myrick, III. 1994 *The Measurement of Environmental and Resource Values.* Resources for the Future, pp. 267.

[2]    "Environment Vise: Law, Compliance" *The National Law Journal*; Corporate Counsel. Monday, August 30, 1993, pp. S1.

[3]    White, Thomas M., Arthur Andersen & Co., *"Measuring Corporate Environmental Performance,"* July/August 1994; *The Environmental Forum*, Environmental Law Institute, Washington, D.C.

[4]    Friedman, Frank B.. 1988 *Practical Guide to Environmental Management.* Environmental Law Institute Monograph, Washington, D.C., pp. 30.

[5]    Kinlaw, Dennis C., 1993 Competitive & Green . *Sustainable Performance in the Environmental Age.* Pfeiffer & Co., San Diego, CA, pp. 279.

[6]    Harrison, E. Bruce, *1993 Going Green: How to Communicate Your Company's Environmental Commitment.* Business One Irwin, Homewood, IL, pp. 57.

# 4

# ORGANIZATION OF A MANAGEMENT PROGRAM

*Gordon A. West, Pilko & Associates, Inc.*

## INTRODUCTION

The methods companies have used to organize and staff their environmental, health and safety programs have evolved over a period of three decades. Until recently, this evolution has been a series of reactions to changing legal requirements and stakeholder demands. The increasingly apparent impact of environmental issues on market position and profitability, however, has caused companies to look at their programs critically to assess how well they are designed to avoid or minimize these impacts and to control costs.

The great environmental lesson of the last three decades of the twentieth century is that governments come and go and political priorities shift, but the public's demand for a "clean" and "safe" environment holds steady. The life of a marketplace is much longer than that of a Congress or a presidency, so there is a need to establish a course that provides certain long-term support for the conduct of business.

Leaping from starboard to port to counter the irregular waves and winds of regulatory enforcement is a strategy less rewarding than one that charts a course to more tranquil waters where enforcement is a lesser concern. Today, leading companies are organizing to be proactive and to prevent problems from occurring in the first place.

To understand how companies are organizing their environmental management programs and what changes are being made, the programs of seven major corporations were reviewed. The companies studied are all multimillion dollar sales, multi-business, global companies. Their businesses include aerospace; chemicals; automotive; oil, paper, and wood products; pharmaceutical; consumer and telecommunications products and services. Their operations and products vary but they all have significant potential impacts on health and the environment. Their programs are mature ones that require and have been given major staff and financial resources.

Companies whose operations have less potential impact on the environment and whose organizations are smaller and simpler do not require the same level of EH&S effort. These seven companies however, are obliged to stay at the leading edge of EH&S management to meet their core business goals. Understanding how those who have to be leaders are organizing their EH&S programs and managing the interface with business and facility management will help others to structure their programs appropriately.

The review of these seven companies focused on three basic questions:

1. What is the involvement of top management (board of directors, CEO, and business unit managers) in the EH&S programs?
2. How are staff resources deployed at the corporate, business unit and facility levels to perform EH&S functions?

3. From what sources are EH&S personnel hired, how are they trained, and to what extent are they interchanged with other units of the company?

## INVOLVEMENT OF TOP MANAGEMENT

The level and consistency of the effort that goes into an EH&S program depends largely on the importance of the program to the top management of the company. Absent a commitment of corporate resources and an inclusion of environmental goals in the vision and business objectives of the company, the EH&S program will flounder. The visibility of top management, in person as well as in words, is also essential to ensure that commitments to achieving environmental goals are made down the line.

### Board of Directors

Top management involvement begins with a board of directors that monitors the EH&S management program and the progress that is being made in improving performance. The boards of all seven companies studied are and have been involved in EH&S oversight for at least the past five years. All receive regular, not occasional reports on program performance. Three Boards have assigned EH&S matters to a committee for closer monitoring and follow-tip. Prominent outside advisors have been retained by the boards of three of the companies to advise them directly on environmental matters.

### Chief Executive Officers

As principal policymaker and spokesperson to all stakeholders, and as principal manager of the company's business portfolio, the CEO has the crucial role in making sure that the EH&S program is in accord with the vision and values of the company, is designed to meet the needs of the businesses, and is being adopted and carried out by the businesses.

The CEO's of the seven companies have been involved in their EH&S programs and have been kept informed on issues and performance for a long time. It is clear, however, that the role of the CEO in environmental matters is becoming a more active one. In recent years:

1.  Three of the CEOs have established senior management standing committees to help develop and coordinate policies and strategies.
2.  Three of the CEOs have joined others in the group of seven including EH&S performance in evaluating and rewarding senior managers who report directly to them.
3.  All seven CEOs are involved in regularly reporting on EH&S policies and performance externally and internally.

## Business Unit Managers

The link in the chain of environmental management that is still being forged is the role of business unit management. Historically, a reactive corporate EH&S staff worked directly with individual facilities to help them comply with changing regulatory requirements and to take care of problems as they occurred. Now the challenge is to take care of problems before they happen and to take advantage of the opportunities that changes in technology and marketplace may provide in order to reduce dependency on materials and processes that impact health and the environment.

In many cases, corporate staff is too far away physically and organizationally to be much help to business managers in making such changes. What, then, should be the role of the business unit manager? Should he stay on the sidelines and merely be kept informed of the performance of the facilities within his unit? Or should he step into the chain of command and be responsible for that performance? What resources does he need to carry out this responsibility?

The review of the seven companies suggests very clearly that the business unit manager is assuming program management responsibility for the unit. A majority, but not all, of the seven companies report that within the past five years, business unit managers have established objectives for the unit's EH&S Program, effectively communicated company and unit policies, and included EH&S performance in evaluating and rewarding the performance of managers.

## Corporate Manager of Environmental Health and Safety

It is important that the person in charge of an EH&S program be of sufficient rank and stature to be able to maintain a program that meets the company's changing needs and cause it to be carried out effectively. The EH&S managers of all seven companies have the rank of vice-president. Most, but not all report to the CEO through another executive (the title of that individual varies). In one case, the EH&S manager is a member of the board of directors. In all cases, however, the position is high enough and prominent enough within the organization to permit formal and informal communication with all corporate and business unit managers and to participate in strategic planning and operations management.

## DEPLOYMENT OF ENVIRONMENTAL HEALTH AND SAFETY

### Resources and Responsibility

Environmental managers are not in charge of environmental management—line managers are. When line managers take the advice of environmental managers, understand it and apply it, the result is the achievement of program goals. The organizational problem, therefore, is to put advisory resources in the right places and at the right strengths to provide correct and useful advice in a timely and consistent manner to those who need it. As noted in the first chapter, multi-business companies making different products and serving different markets all over the world present a major challenge to EH&S management. Of the seven companies reviewed, all but one fully conform to this description of a complex organization.

EH&S management is responsible for managing certain functions that are necessary to support line management. These are

1. Monitor and review global, national, state, and local laws and regulations; and voluntary international, national, and industrial standards that may affect or constrain company operation.
2. Prepare and maintain company policies, guidelines, and standards that encompass legal requirements and reflect company values and priorities.
3. Maintain external communications with important stakeholders including governments, investors, communities, news media, and customers.

4. Conduct audits of company operations to determine compliance, assess risks, and confirm adoption of management systems and policies.

5. Manage major projects including site remediation (e.g., Superfund, business acquisition and divestiture studies, and real estate transfer transactions.

6. Develop and maintain recordkeeping and reporting systems and procedures, and prepare and keep such records and reports that are required to document and measure performance and meet regulatory requirements.

7. Develop and conduct training programs that may be required for compliance or to achieve understanding of company policies and expectations.

8. Provide consultation and advice on technical, regulatory, and company policy matters.

9. Conduct environment and health impact assessments and investigate incidents.

In deciding how to assign responsibility for performing these nine functions, three needs have to be satisfied:

1. The services provided by the program have to anticipate and be responsive to the needs of the businesses and facility operations.

2. Specialized service functions need to be organized efficiently to acquire and maintain expertise, and to avoid duplication of effort and gaps in the delivery of services at different levels and among different units of the company.

3. Adequate monitoring of business unit and facility performance must exist to know that program goals are being achieved and that company expectations are being fulfilled.

The EH&S programs of the seven companies were reviewed to determine what responsibilities for EH&S functions are assigned to corporate staff and which ones are delegated or shared with business unit or facility staffs.

## Corporate Responsibilities

There is a general consensus among the seven companies regarding which functions require placement of primary responsibility at the corporate level. For two of the companies, however, the list of corporate primary responsibilities is longer. One of these companies does not have a diverse group of businesses and because it's overseas operations are also limited, EH&S support in virtually all functions can be provided by a central staff. The other company with a highly centralized EH&S staff does have a very diverse and dynamic portfolio of businesses. Nevertheless, this company has deployed limited staff resources and, essentially, only coordinating responsibility to the business units. The business unit coordinators at this company utilize corporate and facility staff on standing and ad hoc committees to develop strategies and programs for the units.

Primary responsibility for the following functions is generally placed with corporate level staff:

1. Monitoring and reviewing legislation, regulation, and voluntary standards;
2. Preparing and maintaining company standards and policies;

3. Maintaining external communication with stakeholders;

4. Conducting audits; and

5. Managing major projects.

## *Monitoring and Reviewing Legislation, Regulation, and Voluntary Standards*

Monitoring and reviewing legislation, regulation, and voluntary standards is the process by which a company acquires an understanding of what is required to comply with emerging global, national, state, and local legal requirements and accepted good management practices.

Companies whose EH&S programs are sophisticated and expensive seek not only to monitor emerging requirements, but to help shape them. This is done by working closely with company personnel responsible for external communication; by direct contact with government agencies, legislators, and their staff; and through participation in business and industry groups with similar interests. This kind of proactive involvement enables the company to obtain early warning of what new requirements are likely to be, assess what their impacts on company operations will be, and take appropriate steps to prepare for them.

Although this monitoring and "lobbying" function is the primary responsibility of corporate staff, there is a significant amount of consultation with and involvement of business unit and facility staff to identify impacts and priorities and to obtain expert help. Global companies rely heavily on overseas personnel to help them to monitor national and local requirements. Companies with multiple facilities in the United States place considerable, and in some cases, primary responsibility on facility staff to keep track of state and local requirements.

Keeping track of state requirements is a particularly bothersome problem for most companies with a number of facilities scattered across the United States. First of all, federal environmental laws place major enforcement responsibility on the states and provide lots of room for states to enact different and more stringent implementation laws. Second, the procedures used by state legislatures and agencies are not always as formal and well documented as those of the federal government and it is much more difficult to follow what is happening and to predict what is going to happen next. Third, facility staff have day-to-day responsibilities to support operations. They may not have the patience or the time to devote to walking the halls of state government, serving on industrial committees, or preparing commentaries on pending laws and rules that may or may not become final. Many elect to float with the tide and hope that other companies are looking out for their best interests.

The occasion of the periodic compliance audit of a facility is the time when the differing requirements of national and state jurisdictions usually come into focus. As discussed in Chapter 14, companies with mature auditing programs have established audit protocols. It is a common practice, just before an audit is conducted, for an audit team to update the protocol by incorporating any relevant changes in requirements that have been identified through database reviews, phone calls to agencies, or internal consultation.

## Preparing and Maintaining Company Standards and Policies

Preparing and maintaining company standards and policies is the process by which a company articulates its values, vision and expectations for EH&S performance. It is the first step in establishing a hierarchy of corporate, business unit, and facility goals against which performance can be measured and evaluated. It is clearly a corporate EH&S staff responsibility to advise and assist senior management in the preparation of corporate strategies, policy statements, performance improvement goals, and the qualitative and quantitative measures to be used in monitoring performance.

A corporation that sets performance goals to reduce emissions or to reduce the incidence of illnesses and injuries by specific percentages will not achieve those goals without consultation with business units and facilities and without their assumption of responsibility to meet them. Consultation is required to make sure that the reduction goals and timetables are realistic and achievable. Assumption of responsibility is required to create a hierarchy of goals and incentives down through the organization to assure that programs and resources are put in place to meet the corporate goals. The practices of the companies studied and the improvements in performance that they have achieved indicate that a high level of communication and coordination exists within their organizations.

## *Maintaining External Communication with Stakeholders*

Maintaining external communication with stakeholders is the process by which a company:

a. Determines and understands the legal, marketplace, and public opinion forces that impinge upon the businesses of the company;
b. Reports on the results of the company's efforts to respond to the forces; and
c. Constructively deals with the formulation of public policy and public opinion.

Concern about the environment and human health has been at, or very near, the top of the general public's concerns about industry for a very long time. Those who have primary responsibility for communicating with various publics or stakeholder groups need to have ready access to EH&S staff to gain understanding of issues and to get up-to-date, accurate information on company programs and activities. The principal stakeholder groups are

***Customers.*** The emerging demand of the marketplace for products and the facilities that manufacture them to meet new standards of environmental performance is the most important and constructive change that is taking place in environmental management. In his book, *A Moment On The Earth,* Gregg Easterbrook comments that, "...there is one thing market economics does infuriatingly well, that is producing lots of whatever it is asked to produce. Now that capitalism increasingly is asked to produce environmental protection, lots is coming."

At this stage, the role of EH&S in responding to market demands is a minor one. Brad Allenby notes in Chapter 13, "Design for The Environment (DFE)," "With DFE, the role of E&S is to initiate and support change, then act as a resource: virtually all DFE programs will be implemented by non-E&S organizations." This assessment is confirmed by the companies studied who report that their corporate and facility staffs have advisory roles in dealing with customer issues or, in some cases, are uninvolved.

There are two reasons for EH&S organizations to be on the side lines. First, traditional EH&S organizations have focused on facility operations and on "end-of-the-pipe" control approaches (*e.g.*, scrubbers and waste treatment plants) to meet compliance requirements. They have been far removed from basic product and process design and have been brought into the picture only to advise and approve on the handling of waste streams and worker exposures. Second, communicating with customers, understanding and anticipating customer needs, and translating these needs into marketing and R&D strategies are business unit, not corporate or facility responsibilities. In most companies there is no established pattern of communication with EH&S professionals by business and marketing personnel, Input, however, from people who understand legal and good management practice requirements will be critical in shaping successful business strategies. This expertise and knowledge of what customer needs and demands are likely to be is needed at the business unit level and should be close at hand if it is to be effective and timely.

*Investors.* Communication with individual and institutional investors and the financial organizations that advise and manage their interests is a corporate responsibility. It is, therefore, logical that, as reported by the seven companies, responsibility for supporting communications with investors regarding environmental matters is placed with the corporate EH&S group.

*Media.* All seven companies place primary responsibility for press relations and communications regarding EH&S matters at the corporate level. Press reporting of environmental events and issues can reach and significantly affect the attitudes and opinions of all stakeholder groups. It is a matter that is treated carefully and thoughtfully by top management.

Corporate EH&S staff works closely with corporate public relations staff in communicating the company's position or comments on environmental matters. Considerable internal communication with business units and facilities is frequently required prior to enunciating a public statement on an environmental issue or event, in order to fully understand all of the implications to the enterprise and to be as accurate as possible.

The press, furthermore, will frequently seek comments from local management rather than from "P.R." spokespersons from a distant headquarters. Indeed, it may be advantageous to a company to have a local manager speak to an issue and demonstrate knowledge and control of a situation. While it is important to provide central direction for media relations, it is also important to be able to respond effectively to environmental inquiries at the business unit and facility levels.

*Government.* Government policies and actions regarding EH&S matters are among the many issues of concern and sometimes conflict in the relationship of government and business. This relationship is a complicated and important one for large, global corporations. Ongoing communications take place at the highest levels of business and government on long-term policy matters. They also take place at lower technical staff levels of local facilities and government agencies on day-to-day compliance issues.

The relationship of industry and government at all levels, when it results in environmental protection, has been built upon mutual respect, credibility, and good faith. Environmental regulations do not prevent environmental problems from occurring. Agencies have never been staffed adequately to run around checking on whether regulations are being met and to identify changes that must be made to achieve compliance.. They use regulations to state what is required, and to punish offenders after a problem has occurred. Regulations only result in environmental control when they are understood and translated by the regulated community into operating practices that work reliably.

The government relations program on EH&S issues has to be orchestrated to achieve three main objectives:

1. Conformity with company values, policies, and EH&S goals,
2. Consistency in representations being made to all national, state, and local governments; and
3. Establishment of competent, personal, and long-term relationships of company staff, at appropriate levels, with government officials.

This orchestration is achieved by all seven companies by placing responsibility for directing the companies' overall government relations program at the corporate level and providing support from the corporate EH&S group. Business units appear to be relatively in uninvolved at this stage, but facilities have a major advisory role and, in some cases, a primary role in dealing with state and local governments—with advice from corporate EH&S and corporate government relations.

***Community.*** Public perception of how well a company manages the environmental and health impacts of its operations affects the saleability of company products and the acceptability of company facilities. Performance has to be good and has to be communicated. As discussed in Chapter 8, the environmental communications program has to be planned to send the right messages to the right audiences, has to be accurate and credible, and has to be monitored to measure its effectiveness. Global and national community relations responsibility is reported by the group to be a corporate assignment with some participation by business units and minimal involvement of facilities. Local community relations, on the other hand, are primarily a local facility responsibility for five of the companies. In two cases, primary responsibility is retained at the corporate level for local as well as global and national community relations.

## *Conduct of Audits*

The conduct of audits of company operations is the process by which facilities and operating units are reviewed to determine compliance with legal requirements and corporate standards. The audit process also includes follow-up procedures to ensure that audit results are reviewed by appropriate levels of management and that action plans are developed to address problems in a timely manner.Chapter 14, "The Role Of Auditing," identifies three levels of auditing:

a. Oversight, or review of management systems, including compliance audit processes;
b. Compliance assurance, or detailed review of compliance status and management system implementation; and
c. Facility self-audits, performed by site management either as a supplement to periodic higher level audits or as the primary measure of compliance.

All seven companies in the study report that primary responsibility for auditing management systems is placed with their corporate EH&S group. In a minority of cases, some consultation with the facilities and business units is reported.

Primary responsibility for the conduct of compliance audits is also placed at the corporate level by a majority, but not all of the seven companies. In one case responsibility for the audit program is shared with business units and facilities. In another case, the compliance audit is conducted by company facilities themselves with, it is assumed, oversight of the program provided by the corporate management systems audit. Consultation regarding the management of the compliance audit is reported by the group to be higher than for the management systems audit.

As noted above, one of the companies reports the use of the self-audit as the only comprehensive compliance audit. The use of self-audits by the other respondents to supplement corporate audits was not explored in the study. Self-audits, perhaps simplified ones, are undoubtedly in common use in companies with formal programs to follow up on past audits and in anticipation of future audits.

## *Special Project Management*

Special project management assignments are made to a centralized EH&S group when the assignments meet one or more of the following criteria:

a.  The project is one of long duration and high cost and business units and facilities do not have the financial and staff resources to handle it.
b.  Potential financial liability is sufficiently high to be a corporate concern.
c.  Specialized skills and experience are required.
d.  Close coordination with other centralized staff, such as legal and financial, is required.

There are two categories of special projects that commonly are managed centrally by large diversified companies like the seven in this study:

- Major remediation projects are managed centrally including those identified as federal or state "Superfund" sites and those owned properties that are being cleaned up at the companies' initiative. These projects require skill and experience in the selection of the most appropriate and cost-effective technologies; selection and oversight of consultants and contractors; communication and negotiation with regulatory authorities; coordination with legal, financial, public relations, and general management; and communication with other private parties that may be involved in the matter. Most companies involved in these projects have concluded, as have all seven in the study, that they require the full-time attention of a core, corporate group that has developed the expertise to manage them effectively.

- The acquisition and divestiture of businesses and the transfer of real property are transactions that can entail sizeable new liabilities. Acquisitions and divestitures, as discussed in Chapter 13, require close coordination of business, technical, and legal staff in the conduct of due diligence investigations. Although the work does not normally take place over a long period of time, it does take concentrated effort for a period of months in a typical deal. Since soil and groundwater contamination are frequently the major potential liabilities, close access to those responsible for handling remediation project work is needed to assess what those liabilities might be and what the likely cost would be to deal with them. As previously discussed, remediation project management is usually carried out by a corporate group.

- Real estate transactions are not as complicated as transactions involving the transfer of operating businesses. They do, however, require examination of public records (federal, state, and local), site investigations, and interviews of persons with knowledge of the property and its history. Industrial property, in particular, has high potential liability —sometimes in excess of the market value of the property. Real estate assessments require skill and experience to be done thoroughly and efficiently.

• Six of the seven companies place primary responsibility for both business and real property transfers in a corporate group. One company places responsibility at the business unit level. This company is in several very different businesses and has apparently concluded that their businesses are better able to assess the issues that are relevant to them.

**Shared Responsibilities**

Responsibility for the following EH&S functions is generally delegated or shared by the corporate group with business units and facilities:

1. Developing and maintaining recordkeeping and reporting systems;
2. Developing and conducting training programs;
3. Providing consultation and advice on technical, regulatory, and company policy matters; and
4. Conducting EH&S impact assessments and investigations.

With very few exceptions, the seven companies report that the corporate EH&S departments retain some involvement with all of these functions.

## *Development and Maintenance of EH&S Records and Reports*

The development and maintenance of EH&S records and reports has two important objectives. The first is to satisfy the myriad federal and state requirements in the U.S. These requirements include overlapping "incident" reports, hazardous waste manifests, individual permit monitoring data, training records, inventory and emission reports required by state and federal "Right-To- Know" laws, and labeling and premarketing notification laws that exist around the world. Satisfying these requirements is not only mind-boggling and costly, but it is also essential to avoid the most frequent findings of violation by enforcement agencies.

The second objective of the recordkeeping and reporting systems is to provide management with current and trend data on EH&S management performance. Historically, EH&S data has been gathered and reported by individual facilities to satisfy the law and the information needs of facility managers. This same information has then been collected by the corporate EH&S group for their own use in assessing compliance status and for summary reporting to general management. Now, however, management needs more than information. The information now must be linked to individual and unit performance, must be consistent in units of measure and format so that it can be totaled to represent corporate performance in reports to stakeholders, and it must be useful in identifying priorities for program improvement. Data that only shows spills, lost-time accidents, violations, and fines does not help to identify the management process changes that have to be made to eliminate these ultimate consequences.

The review of the seven companies shows that the recordkeeping and reporting process reflects the current management needs of the corporation, the business unit, and the facility. Principal input, of course, comes from the facilities with principal responsibility for managing the whole system placed at the corporate level.

Chapter 6 discusses the awesome task of assisting facilities in the handling of reference information and operations data and drawing from these bases the information that is needed by management at different levels. The seven companies report a significant sharing of responsibility for addressing these problems.

## *Developing and Conducting Training Programs*

Developing and conducting training programs is the process by which a company provides instruction to managers and employees to

a. Satisfy regulatory requirements;
b. Establish general awareness of company and individual legal responsibilities;
c. Establish awareness of company policies, procedures, and performance;
d. Provide technical information regarding specific regulations and on specific company operations; and
e. Reinforce or correct operating practices.

The audiences for these programs include virtually all employees at every level of a company that has EH&S concerns. It is, therefore, not surprising that the seven companies report significant involvement of business units and facilities in the development and conduct of training programs. Corporate staff retains a major role in general awareness, company policy, and EH&S professional training, but a much smaller role in compliance training. Compliance training, to be effective, must relate regulatory requirements to specific operations and technologies. This can be accomplished better by those who are responsible for managing them.

## Consultation and Advice on Technical Regulatory and Company Policy Matters

Consultation and advice on technical regulatory and company policy matters is provided at every level of a company to line and staff persons who are directly or indirectly concerned with business strategy and operations. Responsibility for providing advice is shared by staff at the corporate business unit and facility levels with, hopefully, sufficient communication between levels to keep the advice consistent, accurate, and useful.

The communication challenge is a formidable one. Overlapping and sometimes conflicting regulatory requirements are always changing, as are marketplace demands and technologies. Companies are obliged to establish formal methods of communication through regular meetings and briefings, written and electronic memoranda, and reference information services. Many companies use the audit as a means for communicating changing requirements and expectations and for identifying communication problems. Of greatest importance, however, is the need to encourage and facilitate informal "networking" of EH&S personnel so that the specialized knowledge and experience that individuals may have is made as available as possible to all who need it.

## Conduct of EH&S Impact Assessments and Investigations

The conduct of EH&S impact assessments and investigations is a process with two objectives:

a.  The identification of potential environmental and health adverse effects that *might occur* at any point in the life cycle of a product;

b.  The identification of the root causes of any environmental or health adverse effect that *has occurred* as the result of an incident involving a product or a process.

The results of these studies provide the basis for establishing a program to eliminate or minimize the potential for such an occurrence or reoccurrence. These results become objectives for a process safety (Chapter 10) or pollution prevention (Chapter 11) project.

Assessments and investigations are technical project assignments and are frequently multidisciplinary. Responsibility for carrying them out is generally shared by the corporate, business unit, and facility staff of the seven companies.

**SUMMARY**

The deployment of environmental, health and safety resources and responsibility by the seven companies is summarized in Figure 2.

## *Figure 2*
## FUNCTIONAL RESPONSIBILITY SURVEY
## SUMMARY
[Noted as Primary (P), Advisory (A), Shared (S), or Uninvolved (U)]

| FUNCTION | Corporate | | | | Business Unit | | | | Facility | | | |
|---|---|---|---|---|---|---|---|---|---|---|---|---|
| | P | A | S | U | P | A | S | U | P | A | S | U |
| **MONITOR & REVIEW:** | | | | | | | | | | | | |
| Federal U.S. Legislation & Regs. | 5 | 1 | 1 | | 1 | 2 | 1 | 3 | 3 | 1 | | 3 |
| State U.S. Legislation & Regs. | 2 | 4 | 1 | | | 2 | 2 | 3 | 4 | | 1 | 2 |
| Global Legislation & Regs. | 6 | | 1 | | 3 | 1 | | 3 | 2 | 1 | | 4 |
| Voluntary Standards | 6 | 1 | | | 1 | 3 | | 3 | 5 | | | 2 |
| **PREPARE & MAINTAIN:** | | | | | | | | | | | | |
| Compliance Policies/Procedures | 4 | | 3 | | 1 | 3 | 3 | | 2 | 3 | 2 | |
| Operating Standards & Practices | 2 | 2 | 3 | | 2 | | 3 | 2 | 2 | 3 | 2 | |
| **EXTERNAL COMMUNICATION:** | | | | | | | | | | | | |
| Customers | 2 | 4 | | 1 | 5 | 1 | | 1 | 4 | | | 3 |
| Investors | 7 | | | 2 | 5 | | | 2 | 2 | | | 5 |
| Media | 7 | | | | 6 | | 1 | | 5 | | | 2 |
| Government - National | 7 | | | | 4 | | 3 | | 2 | | | 5 |
| Government - State & Local | 3 | 1 | 2 | 1 | 1 | 2 | | 5 | 2 | 1 | 2 | 2 |
| Community - Global | 6 | | 1 | | 3 | 1 | | 3 | 2 | | | 5 |

*Figure 2 (continued)*
## FUNCTIONAL RESPONSIBILITY SURVEY
## SUMMARY

*[Noted as Primary (P), Advisory (A), Shared (S), or Uninvolved (U)]*

| FUNCTION | Corporate | | | | Business Unit | | | | Facility | | | |
|---|---|---|---|---|---|---|---|---|---|---|---|---|
| | P | A | S | U | P | A | S | U | P | A | S | U |
| Community - National | 6 | | 1 | | | 3 | 1 | 3 | | 2 | | 5 |
| Community - Local | 2 | 3 | | 2 | | 4 | | 3 | 5 | 1 | | 1 |
| **CONDUCT AUDITS:** | | | | | | | | | | | | |
| Management Systems | 7 | | | | | 2 | | 5 | | 3 | | 4 |
| Compliance | 5 | 1 | 1 | | | 3 | 1 | 3 | 1 | 3 | 1 | 2 |
| **PROJECT MANAGEMENT:** | | | | | | | | | | | | |
| Major Remediation | 5 | | 2 | | | 3 | 2 | 2 | | 3 | | 4 |
| Acquisition & Divestiture | 6 | | | 1 | 1 | 5 | | 1 | | 3 | | 4 |
| Real Estate Transactions | 6 | | | 1 | 1 | 5 | | 1 | | 5 | | 2 |
| **RECORDKEEPING & REPORTING SYSTEMS** | 3 | | 3 | 1 | | 1 | 3 | 3 | 1 | 3 | 3 | |
| **DEVELOP & PROVIDE TRAINING:** | | | | | | | | | | | | |
| General Awareness | 3 | | 4 | | | 3 | 3 | 1 | | 4 | 3 | |
| Required for Compliance | 1 | 4 | 2 | | | 3 | 3 | 1 | 3 | 2 | 2 | |
| EH&S Professional | 4 | | 3 | | | 4 | 2 | 1 | | 4 | 2 | 1 |
| Company Policies & Standards | 4 | | 3 | | | 4 | 2 | 1 | | 4 | 2 | 1 |

*Figure 2 (continued)*
## FUNCTIONAL RESPONSIBILITY SURVEY
## SUMMARY

[Noted as Primary (P), Advisory (A), Shared (S), or Uninvolved (U)]

| FUNCTION | Corporate | | | | Business Unit | | | | Facility | | | |
|---|---|---|---|---|---|---|---|---|---|---|---|---|
| | P | A | S | U | P | A | S | U | P | A | S | U |
| PROVIDE TECHNICAL SUPPORT: | | | | | | | | | | | | |
| Compliance | 1 | 2 | 4 | | | 2 | 4 | 1 | 2 | 1 | 4 | |
| Pollution Prevention | 2 | 1 | 2 | 2 | 2 | 2 | 2 | 1 | | 4 | 2 | 1 |
| ENVIRONMENT & HEALTH ASSESSMENT: | | | | | | | | | | | | |
| Raw Materials | 1 | 2 | 4 | | 2 | | 4 | 1 | | 2 | 4 | 1 |
| Manufacturing Process | 1 | 3 | 3 | | 1 | 2 | 3 | 1 | 2 | 1 | 3 | 1 |
| Distribution/Transportation | 3 | 2 | 2 | | 1 | 2 | 3 | 1 | | 3 | 3 | 1 |
| Product Usage | 1 | 4 | 2 | | 3 | | 3 | 1 | | 2 | 3 | 2 |
| TOTAL | 118 | 35 | 48 | 9 | 20 | 79 | 49 | 62 | 21 | 80 | 43 | 66 |

Three general conclusions can be drawn from the survey:

1. Corporate EH&S groups have retained primary responsibility for establishing company requirements and expectations and monitoring operations to confirm that they are being achieves;
2. Making sure that stakeholders understand corporate EH&S objectives and achievements remain a corporate responsibility; and
3. Day-to-day compliance and pollution are responsibilities shared with the EH&S staff of business units and facilities.

## STAFFING AND TRAINING

Environmental, health and safety staffing has been quite stable for the companies studied for the past five years and is projected to remain so for the next five years. There were two exceptions. One company reported a twenty percent reduction in EH&S staffing at the business unit and facility levels (about in line with total employee reductions by the company) and predicted a further twenty percent reduction in the next five years. This company's corporate staff has been and is expected to remain at the same level. Another company reported a five-fold increase in EH&S staffing in the past five years, reflecting a major corporate restructuring.

The EH&S function involves myriad technical professions including law, medicine, safety, fire protection, industrial hygiene, toxicology, epidemiology, biology, geology, meteorology, and a number of engineering specialties. No company can afford to employ all the specialists that might be needed at any one time, so the use of consultants is very much a standard practice. Consultants provide needed skills and help handle major projects and other surges in work load. It is important, however, that company staff be able to provide whatever technical advice is required on a day-to-day basis, and to be able to anticipate problems before they occur.

The experience and skill requirements for EH&S professionals are shown in Figure #3. Building a staff that meets those requirements is a challenge. The most difficult aspect of the challenge is balancing the need to have people who are up-to-date on the rapidly changing technologies, standards, and regulations that pertain to their specialties, and the need to have people who are thoroughly familiar with rapidly changing products and processes of the business.

## ENVIRONMENTAL, HEALTH & SAFETY SKILL AND EXPERIENCE REQUIREMENTS

*Figure 3*

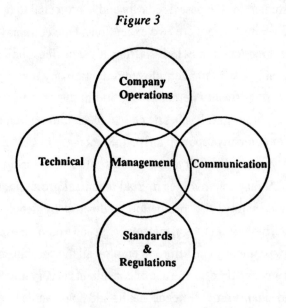

What is the source of new hires into the EH&S function? Five of the seven companies report that new people came into EH&S either from elsewhere in the company, with operating experience, or from outside the company with relevant EH&S work experience. One company reported that new hires came exclusively from within the company. One company—the company that is rapidly expanding its staff—reported that its expansion is being achieved by hiring experienced people from outside the company. They project, however, that future new hires will include people with company operating experience.

How is training provided to the EH&S staff? The seven companies were asked to characterize their training programs:

- Two companies reported that they rely exclusively on internal, formal training.
- Two companies reported that they provide internal, formal training and also use outside training programs.
- Two companies combine on-the-job training with frequent use of outside programs.
- One company reported exclusive reliance on on-the-job training.

It would seem that a deliberate and sustained program to interchange some personnel among corporate, business unit, and facility EH&S departments would promote internal networking and facilitate the development of coordinated programs and goals. This concept, however, has not been broadly implemented. The seven companies reported that the transfer of personnel among levels of the company is

- Deliberate and sustained - 2

- Occasional - 4

- Rare - 1

It would also seem that the transfer of personnel between EH&S and non-EH&S departments and operating units would make good career development sense and also foster a better understanding of EH&S goals by other functions.

The answers were very scattered to the query as to whether such a policy was a "deliberate and sustained" component of the company's management development program. The answers were

- No - 1

- Not but desirable at the
  professional level .............. 1

- Yes, at the professional level
  and desirable at the supervisor level ... 1

- Yes, at the professional and supervisor levels and desirable at the
  department head level ................... 1

- Yes, at all levels ...................... 1

- Desirable at all levels ................. 2

**CONCLUSION**

The programs of the seven companies included in this study show a heavy commitment to reducing the environmental and health impact of their operations. Their environmental, health and safety programs are becoming more closely integrated into the business strategies and EH&S performance has become an important component of total corporate performance.

Preparation for seeking ISO 14000 compliance will bring more refinements to the programs of these companies and others that elect to meet the standard when it becomes final. All of the seven companies have been following the development of the standard—several are participating directly in its development, and several are already taking steps to make changes that will be required. The standard will demand more than compliance with laws and control of environmental aspects within regulatory limits. It will require that environmental management systems and resources be in place, and documented to be in place at all levels of a company where decisions are being made that will affect how company operations and products will impact health and the environment. Whether a company elects to seek third party registration or "self-declare" compliance, documented environmental management systems that integrate EH&S with business operations will be required to assure compliance with the standard.

# 5

# Three Elements Of A Successful Environmental Education And Training Program

*Thomas F. P. Sullivan, Government Institutes, Inc.*

## ENVIRONMENTAL COMPLIANCE REQUIRES EDUCATION AND TRAINING

Almost everyone in business and government is for environmental protection and better occupational health and safety. The average person truly wishes to comply with all the laws and regulations that have been promulgated.

Then why does the nation have so many environmental problems and violations of the law? We believe a primary cause is lack of proper training and education. In fact, most environmental enforcement actions are based on mistakes or "accidents," rather than deliberate violations. This leads to the logical conclusion that there is a need to provide information on the environmental and health consequences of environmental management in order to motivate everyone to do a good job. And there is a need to properly inform all those involved regarding the requirements of all of the major environmental laws and regulations, in order to ensure compliance.

Almost everyone is familiar with the employee who causes problems because he is not properly trained. How many employees *really* understand and are able to do everything that is expected of them in their job? How many employees are properly trained to do all that is required of them when they begin a job? How many employees are retrained regularly to ensure that they maintain competency and are motivated to excellence? If your organization is involved in an environmental accident, will the prosecutor use this lack of training to prove your criminal negligence?

## WHY THERE IS A LACK OF TRAINING

### Relatively New and Changing Field

Most people involved in environmental management have not had extensive formal education in the field. The environmental field has evolved rapidly in the 70's, 80's, and 90's. The field is relatively new and it is difficult for organizations to respond with appropriate educational programs for topics that change so rapidly.

### Cost

Business and government have not undertaken environmental training as a high priority, because of cost. Most businesses and government agencies do not view education and training as an investment-only as a cost factor. This is not intended as a criticism, but simply a statement of fact: training is one of the most commonly overlooked parts of environmental management because people do not wish to spend time and money on it.

## Low Priority and Failure to Appreciate

Two other reasons for lack of training are the low priority that employees give training, and their failure to appreciate the risks involved. The employees already have a full-time job, so how can they afford to spend 40 hours or more per year training? Employees must learn that this training is valuable to them personally; it can benefit them in their career. Also, they and their bosses must understand that the lack of training could result in major problems arising from noncompliance. They must be aware of the consequences for failing to comply.

## Lack of Planning

Lack of training also results from a simple lack of planning. It takes time, money, and planning to develop a trained staff. Until you have been trained, you don't know what the regulations require. In the meantime, the United States government, state governments, and others around the world have been busy promulgating laws, regulations, procedures, and requirements. Thus, in order to keep up with all the developments in this new field, you must be trained initially, and then continue to regularly update this education.

## THREE ELEMENTS OF A SUCCESSFUL PROGRAM

In order to have a successful environmental education and training program. The following three elements are essential:

(1) Creation of a comprehensive plan;
(2) Management, organizational and financial commitment;
(3) Implementation, evaluation, and revision.

Each of these elements will be briefly described in the following sections.

## (1) CREATION OF A COMPREHENSIVE PLAN

The first element in your successful program must be the creation of a plan. You need a road map to get where you wish to go. The road map must be comprehensive and detailed.

A systematic plan for the training of your personnel is needed. Too often training is viewed as a "one-shot deal," which embraces the belief that a one-time presentation in the form of live instruction, a videotape, or training manual will adequately import the knowledge and skills necessary for employees to comply with the constantly evolving, complex environmental regulatory requirements. The training plan must be a system that documents that the proposed environmental training is purposeful-that the activities are clearly linked to organizational goals and objectives. The training plan involves a lot of homework, in that it must

(A) specify the goals;
(B) assess the needs for training;
(C) identify the training needed.

These three plan components will be described in subsequent sections.

## (A) Goals

You start your plan with a statement of your goals-what you hope to accomplish. A successful environmental training program must educate your personnel to go beyond mere compliance with the governing envirorm-lental laws and regulations. You must comply with all the laws in a manner that your employees can understand while performing their jobs effectively.

Environmental training is a multi-disciplinary effort that incorporates information on your company's policies and procedures. This presents a wonderful opportunity to establish lines of cooperative communication with your employees. A clear company policy with respect to obeying the law, supplemented with education that enables your employees to comply with the spirit of the law, is an excellent means of avoiding liabilities in the environmental field. Well-informed employees can identify problems before they occur. Pollution prevention should be one of the goals of environmental training.

How many employees really understand and can do all that's expected of them on their jobs? How many are properly trained? How many employees are retrained regularly to ensure that they maintain compliance? Training goals and objectives must clearly state what your employees should be able to do or know at the completion of training. The goals should be observable and measurable. When the desired outcomes of training are clearly defined, it is much easier to determine the appropriate training needed.

## (B) Needs Assessment

The second part of a plan is a comprehensive needs assessment. Some questions you should address in your needs analysis are

### • Who should be trained?

Almost every employee should be a candidate for at least some formal annual training or you are wasting your human resources. Disgruntled employees are the major source of information for environmental enforcement actions. It is wise to show your employees your major concern for environmental protection through an aggressive training program.

You need a multi-tiered program:

(1) New Employee-as part of orientation for new employees.
(2) Current Employees-what is needed to do their jobs properly.
(3) Refresher (annually)-to keep current employees up-to-date on legal, regulatory, health, safety, and company developments.
(4) New Position-for those employees who will be moving into new positions that require particular environmental knowledge.

## • What information is needed?

First you need to know which of the 700-plus training requirements are applicable to your organization. This training is required by law, so these requirements are your best starting point. Then you can look at other areas, such as: Do your office employees need information on paper recycling? Do your executives need a TSCA or RCRA or CERCLA briefing?

To do this analysis, you need environmentally knowledgeable experts who can understand your processes and systems to guide your planning, particularly on the regulatorily mandated subjects.

## • What training is needed?

What amount of training is needed for the various groups? One hour for the members of the Board? Three hours for staff executives? Forty hours for new hazardous waste managers? Eight hours for purchasing personnel? Sixteen hours for marketing and public relations?

Many considerations must be weighed in determining the amount of training investment. Familiarity with the training requirements of the various environmental and safety laws is critical to determining whether or not your employees must be trained. You must carefully review the provisions of an array of environmental statutes and regulations, in addition to other health and safety laws and regulations, to determine the minimum training requirements that must be met. This is not an easy task and generally requires the use of lawyers or experts to provide the proper guidance. Government Institutes' *Environmental Health & Safety CFR Training Requirements* book could be a big help because it provides the complete text of the 670 CFR training requirements.

The main programs that specify training requirements of interest are RCRA, SARA, OSHA, and DOT.

## • **What are the main program training requirements?**

**RCRA requirements**. Resource Conservation and Recovery Act (RCRA) regulations (40 CFR 264.16 and 265.16) require that treatment, storage, and disposal facility personnel have expertise in the areas to which they are assigned, in order to reduce the possibility that lack of training might lead to an environmental accident. This requirement even covers *generators* that store hazardous waste on a temporary basis (less than 90 days).

The program must be directed by a person trained in hazardous waste management procedures. The content, schedule, and techniques to be used in training must be described in personnel training records, which must be maintained at the facility until closure. This training must be renewed annually.

The RCRA training program for TSD's will be subject to review by the U.S. Environmental Protection Agency during the permitting process. An outline of the program must be submitted with the Part B permit application and it then becomes part of the final RCRA permit.

Because of the variability in hazardous waste management processes and employee functions at treatment, storage, and disposal facilities, the regulations establish no rigid requirements concerning what courses are required, although appropriate topics are suggested.

**SARA requirements.** Section 126 of the 1986 Superfund amendments (commonly known as SARA) required that OSHA promulgate training requirements for worker protection. These requirements can be found in 29 CFR Part 1910.120. These OSHA requirements cover employees involved in responses under CERCLA, such as cleanup of hazardous waste sites, certain hazardous waste operations conducted under RCRA, emergency response to incidents involving the handling, processing, and transportation of hazardous substances and hazardous waste operations at sites that have been designated by state and local authorities.

**OSHA requirements**. There are over 100 other OSHA standards which contain some training requirements. One standard of particular concern is the Hazard Communication Standard (HAZCOM). See below.

*OSHA hazard communication standard training*. Federal and state laws now require that employees "exposed" to hazardous chemicals during normal working activities be trained concerning the specific hazards, use, and handling procedures associated with those chemicals. The Occupational Safety and Health Administration (OSHA) has promulgated a Hazard Communications Standard and many state and local governments have passed various forms of employee HAZCOM or right-to-know laws.

The OSHA Hazard Commununication Standard (29 CFR 1910.1200) requires employers to obtain and distribute information for each hazardous chemical to employees who may be exposed. The standard now requires nearly all companies to comply with these provisions.

OSHA is very explicit about management's responsibilities to provide exposed workers with appropriate training. Companies are required to keep employees informed about the requirements of the standard, how to interpret material safety data sheets (MSDSs), the locations and availability of the written hazard communication program and the identity of the operations in the workplace in which hazardous chemicals are present. Workers must also be instructed in the use of each hazardous chemical to which they may be exposed in their work area.

The specific training must include

- Methods or observations used to detect the presence or release of each hazardous chemical in the work area;
- Physical and health hazards associated with each hazardous chemical;
- Protective measures or equipment required to reduce the exposure to the hazardous chemical;
- Emergency procedures for each hazardous chemical;
- Details of the written hazard communication program.

The federal standard provides a one-time training requirement. However, if a new chemical is introduced or a new employee begins work in an area, further training must be provided. Because of the number of industrial chemicals and individuals covered by these regulations, the impact on the regulated company can be severe. Proper training in this area can help a company avoid lawsuits based on nuisance and negligence.

**DOT requirements.** The Department of Transportation (DOT) also has regulations regarding training related to the transportation of hazardous materials. The trend to legislatively mandate training requirements is expected to continue in new statutes. Failure to train can involve very expensive compliance penalties, further justifying training and education investments. (See 49 CFR.)

**Other requirements.** In addition, other federal and state statutes concerning protection of the environment and public safety also contain general legal authority for regulatory agencies to require employee training. Some states have established training standards for specific industries or employees engaged in specific activities (e.g.. driver safety to obtain a hazardous waste hauler's license) and private organizations also certify qualifications based on training.

It is advisable to seek assistance from your environmental lawyer on the applicable environmental laws to ensure that all applicable federal, state, and local laws have been identified. You probably need to also include experienced, professional training consultants (not lawyers orenvironmental engineers but environmental trainers) in the development of your training plan because, most often, people just don't have the time in their already busy schedule to create a proper plan by themselves and need to create a team or task force.

By assessing the need for training, you may find that, where a regulatory requirement exists, the regulation itself often spells out what the training program should include. However, identifying those employees who should be targeted for environmental training can best be achieved through an analysis of job descriptions, job locations, actual and potential safety and health hazards associated with the job, and the tools, machines, and materials used to perform the job and processes, operations, and maintenance activities. Assessing what each employee is expected to know or do and in what ways the employee's performance or knowledge is deficient will help focus the training on precisely what is needed.

**Requirements beyond compliance?** How can you determine the training necessary for each employee? Several schools of thought exist in determining the training needed.

One school of thought assesses the regulatory requirements and provides training to meet the letter of the law. This approach can miss the intent of regulations and may omit important information for job performance. Another school of thought focuses on providing the barest minimal training, an approach that may not even meet regulatory requirements. This philosophy assumes that "we won't get caught" for not fulfilling our legal requirements for training. However, training violations are the easiest to prove. A third school of thought maintains that there is no such thing as too much training and provides constant training to address all organizational ills. Trainers and managers sometimes think that training is the reason why some organizations are not producing up to par. Recognize that there are a number of management or operational problems that training cannot solve. Training tens of hundreds of employees (for the sake of training) does not constitute effective training.

Effective training focuses on results. It addresses the regulatory requirements and provides the necessary training needed to achieve a specific result, a result that proves to be a benefit to both the employee and the organization. In determining content and quantity of training, you must determine specific training goals for each employee, or category of employees, targeted for training.

The goal should not be mere minimal compliance but satisfaction of the intent, which means you should go beyond compliance.

- ## Is the information changing?

At this time, there are over 1,600 pages of major federal environmental laws and this large body of information is growing. The number of pages of laws, which are the driving force in this field, has more than tripled in the past ten years. This trend is expected to continue as new laws and amendments are added to existing ones.

The regulations implementing these laws have followed the same basic trend line, more than doubling in the past ten years to over 10,000 pages of federal regulations. The changing state laws and regulations are an even more enormous amount of information to comprehend.

Such a large body of changing information needs to be accessed and applied to the regulated community's activities—a most challenging task for those in education and training. We must not only communicate the information, but also try to help managers apply the information in a practical way to their often unique situations plus constantly update it.

**Accurate information and motivation are needed.** The environmental field is complex, cutting across many disciplines-the sciences, health, chemistry, biology, safety, economics, law, engineering, and others. Unfortunately, there are many who claim to be knowledgeable experts in this complex field, but who are not. So, "beware of environmental con-men and con-women."You need accurate information for your education and training programs. Wrong information can be dangerous in a field that involves legal compliance and public health concerns.

Motivation is another key ingredient of successful training. Presenting information alone is not sufficient. You must motivate your employees to retain and apply the information. Since, in today's sophisticated and democratic society, you can't command them to do so, you must give them reasons for, and explain the benefits of, applying that information.

## (C) Identifying the Training Needed

Once we have assessed who needs the training and their informational needs, then we must document the actual training needed. This can involve answering a series of questions:

- ### How is the training to be accomplished?

Can your environmental staff fulfill your training needs or do you need outside expert assistance? Are your trainers experts at training, but not at environmental law? Will you use some engineers that you know, who are entertaining speakers, or a local professor who will speak for a small honorarium; or will you hire a respected environmental training organization to help you?

- ### When should you be training?

Can you train "on-the-clock" or "off-the-clock"? Can you train duringworking hours, after hours, before hours, on Saturdays? How often do you need to do this training? Every quarter, every six months, or annually?

- **What training programs and materials are available that already have been developed?**

Will you use available videotapcs, slidc shows, audio cassettes, pamphlets, demonstrations, lectures? Has your staff researched all the available options before they spend $50,000 making a videotape program?

- **Where should the training take place?**

Can you offer it in your own facilities or is it more cost effective to use other facilities? Should you send employees to public courses out-of-town or train them only locally?

- **What training aids should be used?**

A typical training program generally includes some lectures, supported by overheads, with a pamphlet, textbook, or notebook. Additional audio/visual and other training aids can be prepared to enhance and expand training, but these often require significant resources. Videos have their application, but they can be very expensive if you develop unique ones for limited audience viewing. Slide-tape shows offer cost-benefit trade-offs that should be considered, based on the audience and the application.

One advantage of specially prepared training aids is that they can enhance the site-specificity of the training; site-specificity cannot be overestimated in developing a good training program for all types of training required under the regulations. Site-specific audio/visual aids, although they present an added cost, add greatly to holding group attention and increasing learning retention.

- ## What are the costs versus the benefits?

What benefits do you reasonably expect? What will this training really cost?

You must consider the value of the employee's time plus the cost of the trainer and the training materials. You should weigh the benefits in terms of improved productivity, penalties avoided, and so on.

Based on this information, you can develop a training plan that reflects your requirements in a cost-effective manner.

- ## What is the simplest plan?

A matrix approach is used to describe, as simply as possible, an effective environmental training plan. First, identify who needs to be trained. Separate them into levels as shown on the following list. Then add the specific training hours for each, as shown in Table 1.

## POSSIBLE LEVELS FOR ENVIRONMENTAL TRAINING

**Level I**: Persons who are responsible for overall policymaking in the organization, such as:
- Board of Directors;
- CEO or Company President;
- Executive Vice President;
- Senior Executive Staff (Marketing, PR, Purchasing, Planning, Personnel).

**Level II**: Senior management persons who are directly responsible for the management of day-to-day operations such as:
- Group VP's;

- Plant Managers;
- Division Managers.

**Level III**: Staff persons involved in environmental matters:
- Directors or Managers of Environmental Affairs;
- Staff environmental personnel;
- Attorneys.

**Level IV**: Operations persons whose jobs are supervisory in nature per OSHA and EPA regulations such as:
- Persons who manage hazardous wastes or materials operations.

**Level V**: Field, production, and laboratory personnel who are exposed or potentially exposed to hazardous materials or wastes.

**Level VI**: Office support staff. Table 1 presents a possible Training matrix that could result from your planning efforts.

**Table 1**
**Simplified Sample Annual Training Matrix**

| Topic / Levels | Environ. Briefing | Environ. Laws | RCRA Regs. | CERCLA/ Superfund | Waste Minimi- zation | OSHA | SARA | TOTALS |
|---|---|---|---|---|---|---|---|---|
| Level I | 3 hours | | | | | | | 3 hours |
| Level II | | 16 hours | | | 8 hours | | | 24 hours |
| Level III | | 16 hours | | | 8 hours | 16 hours | | 40 hours |
| Level IV | | | 16 hours | 8 hours | | 8 hours | 16 hours | 48 hours |
| Level V | | | 16 hours | 8 hours | | | 16 hours | 40 hours |
| Level VI | 1 hour | | | | | | | 1 hour |

## (2) MANAGEMENT, ORGANIZATIONAL, AND FINANCIAL COMMITMENT

It is essential that your organization's senior executives, including the CEO, are committeed to training-and to the belief that regular and updated training is a critical component of every environmental compliance program and the key to successful environmental management. In addition, the support of your managers is needed. The managers should have an understanding of the training requirements that are mandated by the environmental regulations and how they impact your organization. This knowledge is beneficial in developing a training plan and securing the resources needed to implement an effective program. Very often you need to "sell" your management. So, your first training is for management. If you don't have their commitment, you probably will never have a good training program. *You* need to work in obtaining this commitment.

You must get a commitment to use resources. This means getting the financial support for the training methods and media selected to implement, evaluate, and improve your training programs; allowing personnel to attend training programs; and committing senior management to assist in the delivery of your training activities. Training is often viewed as an expense, not as an investment-but the failure to comply with the regulations and requirements to train can be very expensive in terms of compliance penalties, injuries, and accidents; damage to property and the environment; and employee morale.

Successful training requires that the entire organization show interest and commitment. Your employees will not gain much from the training (if they even show up for it), if they sense that their bosses aren't interested. You also need financial commitment because good training is expensive. Gaining the necessary financial commitment usually requires the typical environmental manager to do some "selling." You will need to enlist the cooperation of many people-not just top management, but from the top of the organizational structure to the bottom.

Remember, you are competing for limited resources. Be prepared to do a "selling job" with the plan you created, so that you can win the required commitment. You will have to talk about costs. You will have to describe how the benefits far outweigh the costs. In the following sections we will discuss some of the points you can use in "selling" your plan.

## Training Is an Important Investment

A totally different attitude is evolving toward business training and education.Training and education of corporate employees is estimated to cost almost $100 billion annually and over twenty million students are engaged in this corporate education process. Most importantly, such statistics show that America's large corporations are being driven by necessity and the conviction that investment in human resources is just as crucial to their success as capital and plant investments.

Most successful organizations now believe that their employees are their most important asset and willingly invest training resources in those employees. Further evidence of appreciation of the benefits of training is demonstrated by some corporations, which now measure their training expenditures as a percentage of sales and advertise this as an inducement to attract customers and superior employees.

Every environmental manager should implement this trend in his or her organization. Human resources truly are as important as plants and capital investments—probably even more important.Training and education are investments in a company's human resources that can be particularly beneficial in the environmental field because most violations are caused by human error. Training will help to avoid problems before they happen!

## Training Is Mandated by RCRA, SARA, AND OSHA

Another reason to invest in training and education is that it may be required by law. RCRA, SARA and OSHA specifically mandate training.

Failure to train can be disastrous because it may introduce you to EPA's or OSHA's enforcement program. Training violations are easy to prove. You either have the training records and documentation—or you don't. If you don't—you lose!

All corporate and plant managers should be aware of two major developments in environmental enforcement. The most significant trend is the emergence of criminal enforcement as perhaps the most effective deterrent for preventing environmental harm and encouraging sound environmental practices. A second trend, already mentioned above, is the record level of state/federal enforcement actions and the increased level of civil penalties and other related sanctions that are being levied as part of a program to achieve predictable, systematic national enforcement.

The traditional quantitative measures of performance show that both the EPA and state enforcement programs have been operating at record levels. Eacb year criminal fines attain new records, attesting to the increasing emphasis on serious prosecution of violators.

Further, these criminal enforcements are focusing on individuals rather than on corporations. Since most employees realize that a prison sentence is not career enhancing, awareness of these enforcement cases should act as a motivator for learning environmental compliance.

## Our Legal System Presumes Environmental Training

Environmental federal and state prosecutors are focusing on criminal convictions against individuals to enforce environmental laws because government officials have determined that criminal convictions of individuals are a much more significant deterrent than a fine or other penalty.

Some individuals have pleaded that they "just didn't know." Nevertheless, our legal system has developed a concept of "should have known" or a doctrine of constructive knowledge. If the facts warrant. knowledge will be imputed to individuals.

Rather than have knowledge presumed, it is certainly common sense, cost effective, and much better in all respects to have the training to prevent the criminal violation.

## Training Is Good Business

Ignorance is no longer bliss in the environmental law field. Thus, proper environmental training is essential because it is *required* both by law and common sense. In addition, it is one of the *best protections* against costly mistakes that could result in great expense, possible penalties, and even criminal liability. That makes environmental training a key element of good businessmanagement.

Good business management in today's society dictates environmental training. Training is more than just a "good idea": it is a necessity, because the untrained or improperly trained employee is more likely to make serious, even fatal, mistakes that can be very costly in time, money, and resources.

## Training in Lieu of a Penalty for Violations

An emerging concept is to have a violator conduct training instead of paying a penalty. This innovative settlement provision focuses on preventing future problems as well as correcting existing violations. In these settlements, training has been made an integral part of the arrangements. The violator is required to spend money on training in lieu of a fine, which really benefits the violator and the public much more than only a fine.

## (3) IMPLEMENTATION, EVALUATION, AND REVISION

Once you have your plan and the necessary commitments, you are ready to implement your program. The key to proper implementation is taking a positive approach.

If you tell the students they must attend, you will get rather passive acceptance at best. However, if you explain the benefits of the program and make it an enjoyable and beneficial experience, you will receive enthusiastic support.

A critical part of any education or training is objective evaluation. You must get feedback from the students. This feedback from the evaluations may just be a written survey form. Usually simple grades or boxes to be checked lend themselves best to compiling results.

These evaluation results must be analyzed so you can improve. You look for trends, not exceptions. You don't wish to gear your endeavors for one student but the majority. Criticism from one vocal student (who knows it all) should not be taken too seriously. Look for how you can improve the program on a continuing basis because we are always learning and the organizations committed to learning will be the successful ones in the future.

## Plan the Time to Do It Yourself or Buy the Time from Others

Most people don't plan the time for the necessary training or, if they do plan the time, more immediate needs take priority and the training is postponed. To operate your business efficiently, to avoid penalties, legal fees, and organizational problems, you need a strategic training and education program. If you don't have a good training and education program, create one now or pay someone to do it for you.

You can hire training experts to create your plan and do your training for you. But you must obtain the commitment to provide the proper level of *both* monetary and organizational support for the training effort to be successful.

Never underestimate the role that your employees play in how your organization is perceived by the public. They are the formal and informal ambassadors—in contact with the community every day. If a company's employees, through training, are confident in product quality and corporate commitment to safe and environmentally sound operations, community and consumer confidence will follow. The work you do and the training programs you establish for your organization have the ability to affect both employee and community perception.

Be aware that U.S. Department of Justice officials estimate that over 70 percent of environmental violations are initiated by employees' complaints. For those of you responsible for training your fellow employees, your role is now more pivotal and important than ever. And it will continue to change and evolve as the 1990s progress—just as our environmental priorities have.

\* \* \*

For further readings see *Environmental, Health & Safety Training Requirements,* published by Government Institutes.

# 6

---

# INFORMATION MANAGEMENT

*John W. Coryell, Ph.D. and Gary Unruh, AT&T*

## INTRODUCTION

In our multi-media information age, data bombards us from many sources, often unscreened for its value in assisting us to meet our personal and business objectives. Screening of this data to extract what we need consumes a significant portion of our valuable time.

In the EH&S function, the data storm is of Hurricane Andrew proportions. The data set of federal environmental regulations (40 CFR) alone numbers over 14,000 pages and seems to increase weekly. In addition, at the federal level, OSHA, DOT, FDA and other agencies issue regulations that impact the EH&S function and businesses. And, the states and some local municipalities issue their own EH&S regulations governing a business's activities. Corporations set goals and targets that require data collection to monitor progress. Other measurements are made, and data collected, to comply with regulations and control processes which have an EH&S impact.

The challenge to an EH&S function is to accumulate the data, aggregate it into information, and synthesize it into knowledge that can be translated into decision and action. This is not an easy task.

Technologies are increasingly being applied to greatly facilitate the twin tasks of staying on top of environmental regulations and trends and maintaining appropriate records and reports. These technologies are great timesavers, when used correctly. But more significantly, they allow the time and expertise of the environmental professional to be leveraged; he does not have to know the regulations and records verbatim, he merely has to know how to access them quickly. He can then concentrate on the application of the regulations and spend less time and effort as a clerk.

Conoco president and CEO Constantine S. "Dino" Nicandros stated in an address titled *Environmental Management - Burden or Opportunity*, that "better environmental management is, indeed, an opportunity, as it can enhance productivity, upgrade the quality of our operations and add a new dimension of commercial opportunities within our industry."

In his address, Nicandros discussed three areas: the creation of a company culture with a good environment as one of its top priorities; the need for a strategic focus in addressing environmental improvement and the engineer's evolving role in this effort; *and the need for management systems to monitor progress once environmental momentum is under way within an organization.*

This third element becomes a mechanism to accomplish continuous improvement of an organization's environmental performance. Metrics are necessary to assess performance against goals in order to effectively manage the function.

Obviously, environmental management is not solely the responsibility of the environmental manager. People at all levels of an organization must be committed to compliance and environmental performance improvement.

Different information is necessary at different levels, thus information management needs differ. Information systems, whether paper or electronic, must be able to meet the needs of each level if they are to be effective. Plant level operators and environmental professionals require much more detailed information regarding the effect of their operations on the environment than does a corporate level environmental manager who is measuring performance companywide.

## INFORMATION TO BE MANAGED

Prior to 1965, there were few federal laws which governed environmental behavior. However, the grass roots environmental movement of the 1960s caused an awakening at the federal level and legislative activity increased dramatically in the 1970s and continued at a quickened pace through the 1980s. The 1990s is continuing this trend with the promulgation of the Clean Air Act, Pollution Prevention Act and Oil Pollution Act, and upcoming RCRA, Safe Drinking Water, and CERCLA Reauthorizations. Interpreting, influencing, and responding to the effects of new legislation and resulting regulation is not a simple task for the environmental manager and his staff. The volumes of information which must be reviewed and interpreted is staggering. This information must be properly managed if it is to be of value, particularly in this era of corporate restructuring, re-engineering and "rightsizing."

Resulting from these regulations are stringent recordkeeping and reporting requirements which affect all levels of the corporation. Discharge monitoring reports, air monitoring records and reports, PCB annual reports, manifest recordkeeping and reporting, tank testing records, Title 313 Form R, audit records and training records are just some of the compliance mechanisms we are faced with today. In order to comply with regulations, corporations must use data collection systems to help manage information about their operations. Company initiatives such as waste reduction goals add to the demand for effective data management systems. The general public is continually demanding more information about an operation's effect on the environment. Some of the more specific categories of information to be managed include:

| | |
|---|---|
| Environmental Incidents | Best Practices |
| Fines & Penalties | Compliance Calendars |
| Emissions (air, land, & water) | Asbestos Management |
| Chemical Inventories | PCB's |
| Material Safety Data Sheets | Hazardous Material |
| Training Requirements & | Transportation |
| Recordkeeping Audits | Federal and State Regulations |
| Company and Industry Resources | Environmental Costs |
| Occupational Injuries and Illnesses | Toxicological Data |

## INFORMATION MANAGEMENT

The term "information management" encapsulates two related but separate activities: management of reference information and data management. Reference information management involves careful selection and presentation of the result of the information age—a deluge of regulatory, industry, and company reference information. Data management involves managing the environmental data generated by a company's business activities.

# REFERENCE INFORMATION MANAGEMENT

In today's electronic age of computer dial-up information services and availability of compact data disc libraries, reference information can become unmanageable rather quickly as information specialists scramble to meet the growing needs of environmental managers and professionals. Often, organizations gather more and more information without considering how it needs to be filtered for effective utilization. Digesting this information creates apprehension and fear of the very information provided to help us do our jobs. Therefore, information management is critical to present applicable information to appropriate audiences.

Mobil came to this realization in trying to deal with the deluge of regulatory and legislative information sources now available. No single information source was available on the open market which would provide safety, health, and environmental reference information on federal and state regulations, and current legislation in the way Mobil needed it.Those needing to remain current on this information were forced to consult multiple computer systems to gain answers to seemingly simple questions. Mobil decided to take things into their own hands and negotiate licenses with the information vendors to allow them to consolidate five such information sources into a single database, easily searchable by safety, health, and environmental professionals. The Mobil system is now being marketed.

In Dupont, an abstracting service has been set up for environmental professionals. Weekly, brief summaries of significant legislative, regulatory, or judicial actions are electronically sent out to environmental professionals throughout the company. Those who want detailed text for a particular topic can request it, either electronically or in hardcopy.

Electronic reference information sources are abundant. See Government Institutes' *Environmental Guide to the Internet, Eco-Data: 1995 Edition Using Your PC to Obtain Free Environmental Information,* and *Directory of Environmental Information Sources, 5th Edition.*

While reference information is increasingly available electronically, hardcopy sources of information may still need to be utilized to augment these sources. In many cases, the same information is available in hardcopy form and in electronic form, allowing environmental professionals access to the published information now, with improved electronic access later, once momentum to computerize increases. See Government Institutes' *Directory of Environmental Information Sources* for further listings of relevant environmental information.

## Data Management

People, from maintenance mechanics and plant operators to the CEO within an organization, and from regulatory agencies and trade associations to the general public, have a need for factual data about the company's environmental performance. Environmental managers are asked from all directions to provide accurate and timely information and systems to satisfy the demands of these constituents. They all have a need for information about operations, but in different forms and levels of detail.

The concept of data roll-up becomes a necessity, whereby details are collected at the operations level and increasingly summarized as it moves up the corporation. This seems simple enough, but what if all the company's operations are not initially collecting the information at the same level? What if different information is being collected by different operations? And, what if different data base management systems are being used on differing computing platforms?

An information and data management strategy is needed. This strategy should address the data that needs to be collected, the computing direction, such as operating platforms (mainframe, personal computer) to be utilized, and the need, or lack of need, for standardization of systems across the organization.

## ENVIRONMENTAL DATA MANAGEMENT STRATEGY

The process of recognizing that a corporate data management strategy must exist can be painful due to the number of people in differing businesses who must agree and work together to form such a plan. Also a number of mistakes have probably been made with stand-alone, incompatible systems before the need for a corporate strategy is recognized.

Such a strategy does not necessarily state that all operations use the same software products, but it should ensure that the products will (1) meet operations needs and (2) allow data to be easily moved and rolled into the next level of summary systems. Of course, economies can be gained by centralizing company efforts in the areas of software development, acquisition and support. What is necessary though, is defining where the company is now, where the company needs to be, and what needs to be accomplished to achieve defined goals. As a part of this strategic analysis, it is important to consider which information management needs should be addressed corporately and which are better solved by each site or business unit.

Once strategies for database, computing platforms, and centralized or decentralized systems are established, then acquisition of application systems can commence. The desired result is to have data management systems which meet the needs of multiple layers of different businesses and sites, and provides corporate management with the necessary information.

## NEEDS ASSESSMENT AND REQUIREMENTS DEFINITION

Determining the need for environmental information management systems and determining what such systems should accomplish for the organization is a process not much different than buying or building a house. The similarities are presented in the following table and relate the necessary steps required to accomplish this effectively.

- Recognition of Need/Requirements;
- Feasibility, Solution Evaluation & Design
- Implementation;
- Deployment (or moving in);
- Ongoing Operations & Maintenance.

All the steps are necessary to effectively implement an environmental data management system. Omit any of the steps and you are likely to have a system which does not meet the needs of the company or of the individuals who use the system and information it manages. You could end up with a one-bedroom house for a six-member family!

Due to the variance of needs and complexity of a corporation, a comprehensive needs assessment must be performed prior to selecting a commercial system or embarking upon development of a customized system. A number of environmental data base management systems (DBMS) exist on the open market today. Many are modular in design in that you can implement software for one element of environmental information management—waste manifests, for example—or you can implement many modules for a fully integrated environmental management system.

The availability of these products greatly reduces start-up time when implementing a system. Unfortunately, this can cause some to be too hasty in implementing the product before doing the up-front work necessary to properly define the data to be managed and determine if the system really does what they need it to do. The result is that the computer and systems people get blamed for a system that does not "work." Actually, the system works just fine, it just works differently from what is truly needed.

A crucial element of the needs assessment is to identify how the business is currently being conducted. Determine the information needed to effectively manage the business and how that information is currently managed. It may be found that all the information needed to manage the business is not being supplied or generated by current business practices. Perhaps this is why a computer information system is being sought. With increasing emphasis on environmental management by many companies, this may be a new aspect of the business which hasn!t been well managed before. In this case, it is imperative that the business practices be defined and established and that they drive the system not vice versa. If a centralized effort to address the problem is the approach, an analysis of how each site's or business unit's needs differ should be conducted and, if they are significant, efforts to standardize the business practices from unit to unit will payoff. If it is assumed that all units are similar in practice, and that one unit can define the needs of all units, the system will meet with limited success until such standardization has occurred.

Suppose your family grows, or that the walls are cracking and the ceiling is falling in and you decide you need a new house. You have reached the RECOGNITION OF NEED phase in acquiring a new home for your family. But, what are your REQUIREMENTS for the house? How many bedrooms will you need? Do you want a brick house? How many stories? Basement or slab home? What neighborhood? How much money can you afford to spend? And what are your OBJECTIVES and alternatives based upon these requirements? Should you buy a used house, a new existing house, or should you hire a builder and build from scratch.

You must EVALUATE THE SOLUTIONS and determine the FEASIBILITY of each. Will an existing home meet my requirements? Can I find the right size and features in the desired neighborhood? Can I afford to build or do I need to look for a used home? If deciding on a used home, will it satisfy an acceptable number of my requirements? If building, you must now enter a DESIGN phase. You hire an architect/builder and begin working on the blueprints. Perhaps you model your home after an existing floorplan that is desirable, modifying it somewhat to meet your requirements.

Suppose significant new regulations are imposed, or your management makes the decision to become proactive in environmental management and you decide you need a database management system. You have reached the RECOGNITION OF NEED phase in acquiring a system for your company. But, what are your REQUIREMENTS for the system? What categories of environmental information need to be tracked? From what operations and in what form is pertinent information generated? Who will need access to the information? How do they need to view the information? What computing platform(s) does your company operate on? What is the future use of the data? How much money can you afford to spend? And what are your OBJECTIVES and alternatives based upon these requirements? Should you buy a shrink-wrapped system from a vendor, develop your own spreadsheet or manual tracking system, or should you hire a contractor and build a system from scratch?

You must EVALUATE THE SOLUTIONS and determine the FEASIBILITY of each. Will an existing vendor system meet my requirements? Can I find the right features and functionality on the desired platform? Can I afford to build or do I need to look for a more cost effective solution? If deciding on an existing vendor system, will it satisfy an acceptable number of my requirements? If building, you must now enter a DESIGN phase. Your hire a consulting firm or company information systems department and being working on the design. Perhaps you model your system after an existing system that is desirable, modifying it somewhat to meet your requirements.

Once this front work has all been completed, agreed upon by the entire family, and determined to meet local building code, construction or IMPLEMENTATION of the house may begin according to the blueprints. If you are buying an existing house, you will begin contract negotiations, seek financing, and begin packing and preparing to move.

Once implementation is complete, you may begin DEPLOYMENT or use of your new home. You hire the moving company and arrange your existing belongings in your new home. The builder is contracted to correct any construction problems and everyone in the family is trained on how to operate the appliances. You proceed with the final phase of ONGOING OPERATIONS & MAINTENANCE of the house. You must clean house, make minor repairs, and mow the lawn.

Once this up-front work has all been completed, and agreed upon by the groups, who will be using the system and data, and how to meet regulatory reporting requirements. Construction or IMPLEMENTATION of the system may begin according to the design. If you are buying a vendor system, you will begin contract negotiations, seek financing and collect your data, and prepare it to be entered into the system.

Once implementation is complete, you may begin DEPLOYMENT or use of the system. You begin entering your company-specific data into the system, perhaps including several years of historical data. The contractors who built the system are contacted to correct any system bugs and system users are trained on how to operate the system. You proceed with the final phase of ONGOING OPERATIONS of the system. You must update data, generate reports and correct system bugs.

## COMMERCIAL SYSTEMS

There are literally hundreds of environmental software products available on the open market.They range from sophisticated integrated safety, health, and environmental data management and compliance systems, to more narrow systems designed to assist in solving specific problems. Donley Technology (Box 335, Garrisonville, VA 22463, (703) 659-1954) publishes the *Environmental Software Directory*), a collection of abstracts describing nearly 800 commercial and government systems. This directory is an excellent starting point when researching product vendors.

A number of companies provide consulting services to help you start identifying your needs.These companies are able to build an appropriate system for your company either from scratch or by modifying "shells" of systems they have developed for other clients. If taking this approach, involve in-house systems groups to work with the outside company to ensure that systems to be implemented fit within company computing policies and stated technology directions. This will help enable future interface to other company systems (personnel accounting, process management, and laboratory). Failure to maintain consistency with corporate computing standards could mean an early demise of the environmental data management system

One advantage of utilizing vendor supplied systems is that, for a relatively small startup cost, your sites can begin managing their environmental data while continuing to define their ultimate needs in a system particularly if environmental information management is a new activity for the environmental professionals in a site, or if the function being managed is a new business for the site. After gaining some experience with the information they are managing, personnel will be in a better position to define what a system needs to do for them.

Whatever the choice, to buy or to build a system, don't underestimate the effort to deploy the system. Like moving into a new house, the system is void of any furnishings (data) and may need some additional "finishing touches" (customized reports). The new system win consist of software programs and database structures, but sites will need to collect, organize, and enter data into the system.

## SUMMARY

A great number of information resources and database management systems exist to help the environmental manager and his company keep up with our rapidly changing regulatory environment and to meet recordkeeping requirements. It is a challenge to effectively deliver reference information to the broad audience that needs access to the information, and equally challenging to manage company-generated data via a data management system. Electronic systems will only perform what they are designed and programmed for. It is critical then, that systems are properly designed to meet the needs of various users within the company. An overall environmental data management strategy is necessary to allow databases and applications to converge, rather than diverge in the future. This is key because we don't know how we will be required to use this data in the future. When developing or acquiring specific systems, it is critical that the up-front needs assessment, and requirements and objectives identification be done. It is only by focusing appropriate attention on these up-front activities that a system will be able to meet your business needs and be flexible enough to change along with the business.

# 7

## COST ACCOUNTING

*Richard B. Storey and Dana M. Glorie, Arthur Andersen and Company*

### WHAT IS ENVIRONMENTAL COST ACCOUNTING?

Environmental management costs companies up to 10 percent of their total costs.[1] Even more distressing. The cost of environmental mismanagement can financially ruin otherwise healthy companies. These two facts have caused environmental cost accounting to become a priority in the executive suite. Management needs the answer to questions such as

- How much are we actually spending on environmental management?
- How effective is our environmental management program, given the amount of time we spend?
- Can we invest our money in environmental management more effectively?
- Which products cost the company more to produce, from an environmental management standpoint?
- How can these costs be incorporated in our strategic decision making and product profitability models?

By developing an environmental cost accounting methodology, companies can begin to answer the emphatic questions of their corporate executives. Environmental cost accounting requires that environmental costs be identified and removed from overhead accounts and be assigned appropriately to products–the "objects" of the business that cause the costs to be incurred. This involves identifying, measuring, and analyzing costs–both direct and indirect—that are related to environmental activities. The activity and product cost information that is gleaned from this analysis will assist managers in making corporatewide strategic decisions concerning capital budgeting alternatives, process or product changes, new product introductions, and facility expansions.

## WHY DO COMPANIES NEED ENVIRONMENTAL COST ACCOUNTING?

In recent years, the pace of environmental legislation has quickened, bringing industry the burdens of ever increasing amounts of environmental documentation and recordkeeping, increased capital outlays for pollution control equipment, increased operating expenses associated with environmental activities, and increased investment in research and development for pollution prevention efforts. Today, environmental expenses are sometimes as high as 10 percent of a company's total costs.[2] More important, the fines, litigation, and remediation costs that can result from convictions for violating environmental regulations can literally force economically viable companies into bankruptcy.[3] These concerns, coupled with a general public that has become environmentally vigilant and anxious to punish companies that are perceived as environmentally careless, make identifying and understanding total environmental costs and evaluating the effectiveness of the dollars spent, an important corporate priority.

## WHAT IS CURRENTLY BEING DONE?

Most companies still account for the majority of environmental expenditures by charging them to overhead accounts that are not assigned or attributable to specific products or manufacturing processes. The World Resource Institute (WRI) recently completed a study of 11 companies that focused on how companies handle environmental costs. The study found that the majority of environmental costs are hidden in overhead accounts and that companies are unable to fully understand what is driving the costs that are incurred or to assign accountability for measuring and controlling the costs. (Some companies attributed a *minimal* amount of environmental costs, such as expenses for waste treatment, fines and penalties, and capital expenditures for equipment like incinerators, back to production lines).[4]

A few forerunners, such as: Bristol-Meyers Squibb, DuPont, Dow, Ciba Geigy, and Monsanto are implementing processes that will incorporate environmental decisions into mainstream business strategy and management accounting. Both Bristol-Meyers Squibb and DuPont are incorporating environmental issues into everyday corporate decisions. Bristol-Meyers Squibb has begun to use a lifecycle methodology to identify environmental opportunities throughout all business processes for its products, and DuPont uses cost-benefit analysis to prioritize their environmental expenditures. DuPont stated that they have found that 80 percent of environmental gains come from 20 percent of environmental expenditures.[5] Monsanto is using activity-based costing to identify environmental costs in three categories: capital expenditures, operating expenditures, and R&D expenditures.[6] Some companies like Dow are still researching the use of full cost accounting to identify and to quantify the environmental impact of all raw materials.

While some forward-thinking companies, like those discussed above, are beginning to establish environmental cost accounting methodologies, many companies still have no reliable method for identifying, measuring, and controlling environmental expenses.

Multi-functional teams comprising representatives from industry, academia, consulting, government, and accounting have sprung up to help address the need for environmental cost accounting in industry. One such team, cosponsored by the Institute of Management Accountants (IMA) in December of 1993, was focused on exploring ways to integrate environmental considerations into management accounting and capital budgeting decisionmaking[7]

## HOW DO YOU BEGIN TO DO ENVIRONMENTAL COST ACCOUNTING?

Activity-based costing (ABC) is a methodology that assigns costs to activities in manufacturing and/or managerial processes based upon the resources that they consume. Activity costs are then assigned to the cost objects (i.e., products, customer groups, etc.) based upon the activities that they require. This methodology provides companies with more accurate product and process cost information, and enables companies to manage process costs based upon the activities in a process, not the functions. The information that is produced by implementing ABC is used by companies to perform more accurate resource planning, to test the economic and operational effect of the product or manufacturing process changes, to reduce the occurrence of nonvalue-added activities, and to improve processes while controlling costs.

ABC has been used for some time in manufacturing and, more recently, in service industries to better understand and manage the cost of activities in numerous management and production processes. Companies can use this same technique to identify the costs of all activities that are performed to address environmental issues and then to assign these costs to specific products or processes. Two complications that must be considered in using ABC for environmental cost accounting are

- Indentifying all environmental costs incurred throughout the entire product lifecycle–Potential liability costs for an environmental cleanup is one type of product life cycle cost that is difficult to quantify, and potentially horrendous. It should be weighed carefully, however, in comparison with the costs of waste disposal alternatives, which offer different risks of future liability.
- Identifying all activities performed for environmental management–a management process that is highly fragmented.

There are two levels of analysis needed to build an ABC system:

- Activity analysis, which determines how resources are consumed by activities; and
- Cost object analysis, which determines how activities are demanded by individual product offerings, customer groups, or any other drivers of costs.

## Activity Analysis

Activity analysis consists of three main steps: identifying major activities, identifying the drivers of activities and related volumes, and determining how resources are consumed in performing activities.

### *Identify Major Environmental Activities*

Identifying activities in the environmental management process can be difficult because the process is fragmented; that is, activities are not sequential and continuous and are performed in many departments and locations. For this reason, it is important to focus on the major cost bearing activities rather than spending significant time in identifying every activity that is performed. One way of focusing on major activities is to consider the goals of environmental management for corporations such as compliance, pollution prevention, risk minimization, remediation, and safety, and to determine the major activities that would need to be performed to achieve these goals. The following are some of the activities that might be performed in achieving the above mentioned corporate environmental goals:

- *Compliance*—environmental auditing, collecting data and preparing reports to comply with regulations, tracking, and communication of legislation and regulation, and the cost of fines for noncompliance;
- *Pollution Prevention*—research and development for pollution prevention, creation, and monitoring of waste minimization programs;
- *Risk Minimization*—cost of "Right to Know" training, cost of purchasing and maintaining protective equipment;
- *Remediation*—activities involved in site cleanup, litigation, and management of superfund sites;
- *Process Safety*—cost of worker compensation.

Once this list of major environmental activities is developed, the next step is to determine where in the organization they take place and by whom they are performed. One approach would be to develop a matrix (refer to Exhibit 1) with activities on one axis and responsibility on the other. Interviews should be scheduled with managers and personnel designated responsible for the activities. The goal of the interviews is to understand how employees are spending their time related to environmental management. The 80/20 rule should be applied here in gathering information. That is, the focus should be on determining and gathering information on those environmental management activities that consume 80 percent of employee time.

*Exhibit 1*

**Responsibility Matrix**

Instructions: Please identify where (corporate, business unit, facility) and by whom the activities are performed. If performed by more than one individual, indicate the manager of the group. Also indicate the number of people that perform the activity (#FTEs). Finally, note if responsibility is primary (P), advisory (A) or shared (S).

| Activity* | Corporate | | | Business Unit | | | Facility | | |
|---|---|---|---|---|---|---|---|---|---|
| | | | | | | | | | |
| Monitor and Review: | | | | | | | | | |
| Federal U.S. Leg. & Reg. | | | | | | | | | |
| State U.S. Leg. & Reg. | | | | | | | | | |
| Global Leg. & Reg. | | | | | | | | | |
| Voluntary Standards | | | | | | | | | |
| Prepare & Maintain: | | | | | | | | | |
| Compliance Policies & Procedures | | | | | | | | | |
| Operating Standards & Practices | | | | | | | | | |

| Provide Technical Support: | | | | | | | | | |
|---|---|---|---|---|---|---|---|---|---|
| For Compliance | | | | | | | | | |
| For Pollution Prevention | | | | | | | | | |
| Maintain Recordkeeping / Reporting Systems | | | | | | | | | |
| Performing Recordkeeping For Hazardous Waste | | | | | | | | | |
| Environmental Reporting | | | | | | | | | |

*The above matrix contains examples of environmental activities.

An activity dictionary should be developed for the most important activities related to each corporate environmental goal. With regard to the goal of compliance, for example, obtaining permits, training employees, performing audits, and recordkeeping for waste disposal would be among the activities documented in the activity dictionary, assuming that it was determined through the interviews that these activities consume a significant amount of resources.

## *Identify Activity Drivers and Driver Volumes*

Key drivers for the environmental activities need to be clearly identified. An activity driver is the event or "trigger" that causes an activity to be performed. For example, the activity to obtain permits is "triggered" every time there is a need to get a new permit. Therefore, the activity driver for the obtaining permits activity would be the number of permits that need to be obtained, in a given time period. The more permits that need to be obtained the more total time employees will spend in obtaining them.

Determining the drivers for some activities will not be as clear cut. For example, it could be determined that the activities for employee training are triggered by the hiring of new employees or by the need to periodically train existing employees. In cases where activities seem to have more than one driver, the activity might be split, in this case into training for new employees and training for existing employees. However, before splitting major activities into smaller ones, the size of the original activity should be considered. If the original activity does not consume a significant amount of resources (e.g., employee time, capital resources, etc.) it may be sufficient to choose the most representative of the drivers identified.

Once the drivers for activities have been determined, the sources for the volumes of these drivers need to be identified. Sources of needed activity driver information include training records, hazardous waste recordkeeping systems, audit records, and project logs, etc. If volume information is not currently being collected on a continuous basis for some of the activity drivers, substitute drivers should be considered. In cases where no substitutes are available, estimates or short-term tally information can be used to build the initial ABC environmental cost system, but procedures should be implemented to begin collecting activity driver information on a continuous basis.

*Exhibit 2*

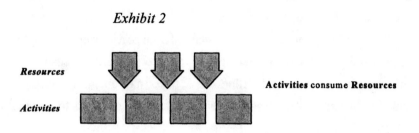

Resources

Activities consume **Resources**

Activities

## *Determine How Resources Are Consumed by Activities (refer to exhibit 2)*

Four types of resources will have to be identified and assigned to activities: payroll expenditures, operating costs, capital expenditures, and current and future liability costs.

Payroll expenses should be assigned based upon the activities on which employees spend their time. Most companies do not have detailed time-tracking systems and defined activities will usually not be identified. Therefore, after the activity dictionaries are complete, the employees that perform the activities must document the amount of time they spend on each activity per activity driver. In other words, how much time does it take to obtain a permit. One way to gather this information is to create standardized activity effort worksheets to capture activity time information. This information will provide the basis for calculating factors used to assign payroll expenses to environmental activities.

The existence of operating expenses and capital expenses should be identified during the interview process. These expenses will then be quantified by reviewing the general ledger and capital budgets. Companies that use the same accounting systems at the corporate, business unit, and facility levels and across business units will have an easier time identifying and quantifying total environmental expenses. They will include items, such as the cost of waste disposal contractors, installation of scrubbers, the installation of new equipment for implementation of a new process that uses water-based solvents, the percentage of worker compensation costs that are attributable to repetitive stress illnesses, or the cost of operating a company-owned disposal facility. Each expense should be assigned to a specific activity.

The last categories of expenses that should be identified are potential liability expenses, such as fines and penalties and litigation costs and remediation costs. The estimation of these future expenses will have to be determined by individuals who have expertise in these matters.

*Exhibit*

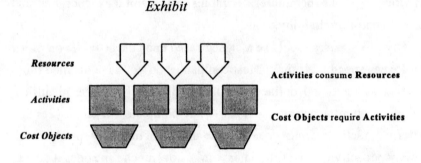

## Cost Object Analysis

Cost object analysis determines how activities are demanded or required by products (refer to Exhibit 3). The relationship between activities and products (cost objects) is determined by the specific activity driver. For example, if the activity driver for disposing of hazardous waste is the number of tons of hazardous waste, each product that produces hazardous waste would be assigned a portion of the cost of the hazardous waste disposal activity. This assignment would be based upon the amount of hazardous waste produced by the manufacturing process for each product. Therefore, the first step in performing cost object analysis is to take an inventory of the activities that have been costed in the activity analysis phase and determine for each product if producing the product requires the activity.

Once these relationships have been drawn, the amount of each activity that is attributable to each product–based upon the volume of a product produced in the period being studied–must be determined. Activities can be grouped into four categories, depending upon the type of event that drive the activity to take place. These categories assist in developing the correct approach to assigning the appropriate cost to individual products or other cost objects.

- Unit-Based Activities —These activities are performed on each unit of a product and will be assigned to products that require the activity, based upon the number of units of the product that were produced in the period being studied. An example of a unit-based activity might be disposing of a "dirty" solvent that is used to clean electrical parts at an electronics manufacturer.

- Batch-Based Activities—An example of a batch-based activity might be the disposal of cleaning supplies used in cleaning production areas between batch runs, and the disposal of solution used in a solvent bath that is used to clean an entire batch of the product. The cost of waste treatment of the discharge from a particular process should be charged to that process and to a product during the period of time it is being produced. The cost should not be hidden in a plant overhead line item called "waste treatment." Batch-based activities, like unit-based activities, will be assigned to products that require them, based on a batch volume basis. Therefore, products that have large batch sizes will not receive a disproportionate amount of cost.

- Product-Sustaining Activities—These activities might include obtaining permits if the toxic releases from the manufacture of one product or family of products required a facility to obtain a specific type of permit or the cost of reporting on the disposal of hazardous waste that is only produced by one product. Activities that are product sustaining would be wholly assigned to only the product that requires it.

- Facility-Sustaining Activities—Examples could include testing and monitoring all the possible sources of air emissions in a facility, or tracking and communicating changes in federal and state regulations. The cost of facility-sustaining activities should be divided among all product lines.

## WHAT ARE THE BENEFITS OF DOING ENVIRONMENTAL COST ACCOUNTING?

The main objective of activity-based costing is to facilitate the management of the cost of activities. This is the first benefit that building an environmental ABC system will provide to companies; the ability to understand what they are spending on environmental initiatives and to better manage these costs.

The outputs of activity analysis, the total cost of each activity, and the cost of an activity per driver will provide companies with much needed cost performance measures for environmental management. These cost measures serve as a baseline against which future improvements can be measured. Companies can review the total cost for each activity in order to pinpoint environmental activities that are the most costly and can focus improvement efforts on them. Next, they can determine whether the strategy for cost reduction should be to reduce the number of times that an activity is performed or to focus on reducing the time it takes to perform the activity.

In managing the cost of activities, however, companies should be cognizant of the fact that cost performance measures should be utilized in conjunction with measures of an activity's effectiveness (i.e., quality and timeliness). The combination of cost, quality, and timeliness measures will provide greater insight into activity performance, and will give a basis for internal benchmarking among facilities and external benchmarking with other companies.

Another benefit that building an Environmental ABC system will provide to companies is a more accurate view of overall product profitability. Companies traditionally compute product profitability with little attention given to environmental costs, even though they sometimes account for more than 10 percent of the total expenses in a company. Also, since potential liability costs will have been factored in for products that are environmentally more risky, a longer term view of product profitability will have been established. Additionally, differences in product costs can be tracked back to their requirement for specific environmental activities, giving corporate management a "roadmap" for improving a product's profitability by reducing its need for environmental activities.

Most important, the information culled from activity and product environmental cost information can be used by companies as a key input to corporatewide strategic decision making such as new product introductions, product or process change alternatives, facility expansions, acquisitions, and capital budgeting alternatives. Modeling operational changes in the Environmental ABC system and analyzing resulting effects on the existing cost structure will provide answers to questions such as

- How will this product or process change affect the profitability of existing products?
- Will it reduce the number of permits that a facility will need?
- Will less time be needed to record and report toxic substance usage?

Similarly, environmental cost data could be used to paint a more realistic picture of a potential new product's profitability or to evaluate the cost of alternatives for capital investments based upon the effect that each investment will have on the cost of environmental activities and ultimately on the environmental costs that products bear.

Although establishing an environmental cost accounting system is no small task, the benefits of incorporating environmental costs into everyday corporate decisionmaking and of having tools to manage the effectiveness of ever increasing environmental expenditures, far outweighs the effort.

## ENDNOTES

[1]    Epstein, Marc J., "A Formal Plan for Environmental Costs."

2    *Ibid.*

3    William, Georgina and Phillips, Thomas, Jr., "Cleaning Up Our Accounting for Environmental Liabilities." *Management Accounting*, February 1994. p. 30.

4    Kirschner, Elisabeth, "Full Cost Accounting For the Environment," *Chemical Week*, March 9, 19994. Pp. 25-26.

5    Epstein, Marc J., "A Formal Plan for Environmental Costs."

6    Kirschner, Elisabeth, "Full Cost Accounting for the Environment," *Chemical Week*, March 9, 1994, pp. 25-26.

7    Freedman, Julian M., "Accounting and Pollution Prevention," *Journal of Management Accounting*, February 1994, p. 31.

# 8

## PREPARING A COMMUNICATION PLAN

*Maria M. Bober Rasmussen and Jeanne E. McDougall, Kodak*

### INTRODUCTION

Effective communications are a key and necessary component of a successful environmental, health and safety program. Internally, the EH&S vision and policies need to be clearly communicated to all employees, so that they are aware of their responsibilities and energized to perform them. Performance needs to be communicated so that employees recognize the impact of their efforts and where more effort is required. And, effective communications allow management to monitor performance and adjust for deviations. Externally, the public wants more information on a company's performance and efforts to improve that performance. Also, customers are increasingly requiring more information on the safety, health and environmental aspects of products. Communicating successfully with these external audiences is of paramount importance so that the good work done inside the company finds a listening ear outside the company.

### COMMUNICATIONS PLANNING

Any good communication program asks similar questions:

- What is the current situation?
- Who is your audience?
- What messages do you want your audience to hear?
- How do you measure your success in delivering those messages?

Environmental communication is no different, but in this area, companies often deliberate on more basic questions:

- How much communicating do you want to do on this topic?
- How high a profile should the company maintain?
- What are the consequences of communicating?

These questions are valid ones. A company may very well have a good record on environmental issues; its policies and programs and management systems may reflect state of the art in environmental management; its processes and facilities may be routinely improved to reduce environmental impacts. But why call attention to your environmental attributes, if doing so places the company into a controversial arena and, quite possibly, sets it up as a target?

The answer is, that not telling the story you have—especially if it is a positive one—is worse than having no story to tell at all. Chances are, if you choose not to talk about your environmental performance, no one else will either—until there is something negative to report. The time to start communicating is not when you have a problem. That is the time to gently remind people that the current situation is an anomaly (remind them of past communications), demonstrate to them that you are appropriately responding to and managing the situation, and assure them that you will learn from the situation and put in safeguards to your processes and systems to prevent recurrence. It is difficult to justify that the incident does not reflect "business as usual" if you have never taken the time to communicate your ongoing programs.

You will hear environmental communications referred to as being either "reactive" or "proactive." The former is often seen as a vestige of traditional corporate reluctance to discuss environmental issues unless absolutely forced to—"don't say anything, and maybe it will go away;" the latter is characterized as a more enlightened way of addressing the situation —a more confident approach, associated with companies that attack their environmental challenges head on, without being paralyzed by uncertainty. If you accept this paradigm, then everyone would strive to be proactive and distance themselves from the unenlightened few who choose to cling to the old reactive approach.

But the fact is that both proactive and reactive communications play important roles in any successful environmental communications program. Creating an atmosphere of open, two-way communication, on an ongoing basis, is essential to building trust and "equity" in the environmental area—or in the words of a public relations representative we know, an "inoculation" against negative stories that may arise. Similarly, the company that is not prepared to react to event-driven changes in circumstance could easily and quickly parlay away any goodwill derived from their proactive communications.

A more accurate categorization of environmental communications would be to refer to them as either "ongoing" or "event driven." In both cases, a strategic approach is essential to keeping your messages consistent, accurate, credible and timely—the key to deploying a successful environmental communications program.

# A STRATEGIC APPROACH

## What is the Current Situation?

Whether you are designing environmental communications as the HAZMAT team is pulling into the parking lot, or as a way to project some long-term messages to an audience as part of your ongoing communication strategy, the first thing to do is evaluate the situation.

Begin with a little history. What environmental communications, generated by you or others, have existed in the past? Include last year's negative headline as well as last month's employee memo. Then list all environmental communications opportunities that you are aware of ("We could do a better job of pollution prevention if employees were more aware of our corporate goals") as well as challenges ("the public thinks our industry is "dirty"). Now look ahead and ask your self and others, "What are our future risks of negative publicity? What are our opportunities for positive publicity and relationship building?" The environmental professionals in your company often know what is coming down the road, including future regulations and facility permitting or re-permitting. Talk to them.

Another way to assess the situation is by thinking of your key publics and comparing their current opinion of your company's environmental efforts with your desired state. Look for the biggest gaps between the current and desired states. Ask yourself, "why does this group have a lower opinion about my company's environmental performance than I would like? What do they need to have a better opinion about us?" The box below provides more detail about using this method.

## *Figure 1*
# STAKEHOLDER OPINION ANALYSIS

**Opinion of Company's Environmental Performance**

| STAKEHOLDER | CURRENT STATE | DESIRED STATE | GAP |
|---|---|---|---|
| Employees | -5 -4 -3 -2 -1 0 1 2 3 4 5 | -5 -4 -3 -2 -1 0 1 2 3 4 5 | |
| Community | -5 -4 -3 -2 -1 0 1 2 3 4 5 | -5 -4 -3 -2 -1 0 1 2 3 4 5 | |
| Investors | -5 -4 -3 -2 -1 0 1 2 3 4 5 | -5 -4 -3 -2 -1 0 1 2 3 4 5 | |
| Customers | -5 -4 -3 -2 -1 0 1 2 3 4 5 | -5 -4 -3 -2 -1 0 1 2 3 4 5 | |
| Activists | -5 -4 -3 -2 -1 0 1 2 3 4 5 | -5 -4 -3 -2 -1 0 1 2 3 4 5 | |
| Media | -5 -4 -3 -2 -1 0 1 2 3 4 5 | -5 -4 -3 -2 -1 0 1 2 3 4 5 | |
| Government | -5 -4 -3 -2 -1 0 1 2 3 4 5 | -5 -4 -3 -2 -1 0 1 2 3 4 5 | |
| Suppliers | -5 -4 -3 -2 -1 0 1 2 3 4 5 | -5 -4 -3 -2 -1 0 1 2 3 4 5 | |

**Process:**

1. With zero denoting ambivalence toward's the company's environmental performance, note your perception of each stakeholder's opinion about your company's environmental performance. Base the rating on what you hear from the stakeholder through conversations, surveys, letters and hearsay.

2. In the DESIRED STATE column indicate how you would like them to feel. You may not necessarily want everyone to be a "5". You may be perfectly happy with some groups being ambivalent (rating a "0") about your environmental performance. Keep in mind that it will take an enormous amount of effort to make everyone a "5".

3. Write the difference between the CURRENT STATE and the DESIRED STATE in the GAP column. For example, if the desired state is "3" and the current state is "-2", the gap is (3)-(-2) = 5

4. Direct your communications efforts towards closing the biggest gaps.

From your assessment of the past and present, as well as your thoughts about the future you can make some major decisions about the significant issues at which you should direct your efforts. And, from there, you can define and then refine your key messages and your key audiences.

## Who is Your Audience?

If your communications strategy is based on affecting the opinions of a specific target group, then you should have a handle on this. But if your strategy is based on communicating about a specific issue, defining your audience is the easiest thing to get wrong, or forget. One communications professional we know has a needlepointed aphorism on his wall that reads, "*You* are not the target audience!"

In any situation are many audiences. The key is determining *who* cares most about the message your developing—and who *you* would most like to hear your messages. And then to be sure to talk to them.

It is important to distinguish between a primary and a secondary audience. Employees and community residents, for example, might be considered primary audiences for a facility-related story; the local media might be considered a secondary audience, because you care about the way they influence your primary audiences. When considering your audiences ask yourself, "do I care more about what this audience thinks, or the way they influence other audiences to think?"

*Figure 2*

## POTENTIAL KEY AUDIENCES

**Employees**　　　　**Community**
**Regulators**　　　　**Customers**
**Suppliers**　　　　　**Stockholders**
**Activists**　　　　　**Media**
**Peer Companies**

Try to keep your list of target audiences for a particular communication brief and specific; it will make your messages more coherent. For example, if you chose employees as your target audience, narrow it down if you can. Do you want to include retirees as well as contractors? Is this communication also for U.S.-based employees or international employees? Do you want to focus on shop floor, sales or, office employees? Senior management?

The better you define your key audience(s) the more targeted the message can be, and the better chance you have of success. If you consider all audiences to be key, then in reality, maybe *none* of them are key.

## What Are Your Messages?

Many communications programs fail simply because their messages are not heard.

Common pitfalls include:

1.　You develop too many messages, no single message is heard;

2.  Your messages don't respond to the questions people are asking themselves;

3.  Your messages answer the questions, but are not consistent with other messages being heard from others in the company, or even from you.

A good way to get at key messages is to pull together a group of your best minds and present them with the situation analysis and the key audiences that you are targeting. Come up with the one or two (three at the absolute most) things you would like to make clear, and then test them on individuals who come from your key audiences.

Before you deploy your messages, review them against other key messages coming out of the company. Your messages must be integrated; there is no surer death than if they are received as a passing fancy, or the "topic of the day." If a key corporate message supports profitable growth, there will be consistency and equity built into a message like, "a new plant —improving the economy *and* our environmental performance." Conversely, if a key company message centers on cost cutting, your message might be "Some things are too costly to cut—worker safety and environmental responsibility." A consistent message is a plausible message.

It is important for any communication plan, but it is even more important for an environmental communications plan. Be accurate and credible. Key messages should be short and to the point. However, you must have information and data to back up what you say. Another important thing to keep in mind—no product or process is perfect, so never claim it is.

## How Do You Measure Your Success?

Before you deploy the first message to your first audience, take an accurate reading of your key audiences' perceptions in areas where you are targeting your efforts. Data collection costs are not insignificant, especially if you use consultants, but if you don't get an "initial read," you will only have anecdotal evidence as to the efficacy of your communications effort.

But this need not be as costly as you fear, or as time consuming. Chances are, others in your company routinely talk to the audiences you have decided are key. Your community relations personnel are in routine contact with your facility neighbors, the sales department has an ongoing relationship with customers, and supervisors and human relations personnel talk to employees. Most, if not all, of these company organizations are probable conducting formal or informal surveys or focus groups to see how well they are communicating to their primary audience. All you need to do is get yourself plugged in to these various sources.

Is it possible to insert a question or two related to your situation? Are there questions on the existing survey that could be impacted by the messages you have developed? If you can't use a pre-existing survey instrument, maybe you can use a piece of their system, even if it is just their mailing list.

As an example, Kodak has an ongoing survey to track employee perception of and satisfaction with the company. The survey consists of over a hundred questions and is sent out quarterly to a statistically significant number of employees. For many years, there has been a question on the company's environmental ethic. Once an environmental communications strategy was devised and employees were determined to be a key audience, the existing survey was reviewed to determine if it would adequately track the success of the communications program among the employees. The conclusion: one environmental question on the existing survey was too broad and would not necessarily convey whether the key message was successfully delivered. Since the owner of the survey was wary about making an already lengthy survey even longer, a new, short environmental survey was devised. The same sampling and data analysis process was used as was employed in the larger survey, and the first question on the purely environmental survey was the one environmental question from the larger survey. Checking to see that we had about the same response to this "overlap" question allowed us to confirm that our representative sample population was the same as theirs.

While using the existing employee survey was not a good fit, in this case, for tracking our success delivering one specific environmental message, the existing survey does provide data on employee feelings about the company's overall environmental performance and is a useful long-term measure of successful communication to employees.

Whatever tools you identify to measure the success of your message, use them. Ultimately, the questions will be, Did you move the needle? Did your communications affect a change? Answering these questions with specificity will help you to design even better, more effective communications the next time around.

## SENDING YOUR MESSAGES

Once you have identified the high-level information—who, what, and why ?—you can focus in on the detail: when, where and how?

A good place to start is with networks; basically, who talks to whom? As mentioned earlier, likely everyone you might identity as a key audience for your environmental communications is already a key audience for someone else in your company. Take some time to learn who is talking to whom, about what, and through what mechanism. Figure 1 provides this information for a fictional company. It is useful to go through this exercise on paper, run it past your contacts in the various communications areas within your company, and confirm that you really understand the current process.

*Figure 3*

## COMMUNICATIONS NETWORKS MATRIX

| STAKEHOLDER | EXISTING NETWORK TO REACH STAKEHOLDER | ROUTINE METHOD OF COMMUNICATION | PRIMARY STAKEHOLDER CONCERNS |
|---|---|---|---|
| Employees | + Employee Communications Dept.<br>+ Supervisor | + Employee Newsletter<br>+ Posters on Bulletin Boards<br>+ Department Meetings<br>+ One-on-One discussions | + Job Security<br>+ Salary<br>+ Working Conditions |
| Community | + Employees<br>+ Community Relations Dept.<br>+ Local Media | + One-on-one conversation<br>+ Community Advisory Panel<br>+ Local Advertising<br>+ Facility Tours<br>+ Editorials/articles<br>+ Nightly News | + Jobs<br>+ Environment<br>+ Property Value<br>+ Personal Safety<br>+ Crime |
| Investors | + Shareholder Relations Dept. | + Annual Meeting<br>+ Annual Report<br>+ Shareholder Newsletter | + Short/Long Term Profits<br>+ Liability/Risks |
| Customers | + Sales Force<br>+ Marketing Communications<br>+ Product Service Dept. | + One-on -one Visits/Service calls<br>+ Product Literature | + Product Quality<br>+ Product Cost |
| Activists | + Public Affairs Dept. | + One-on-one Conversation | + Their Cause<br>+ Membership |
| Media | + Media Relations Dept. | + Interviews<br>+Press Releases | + Readership/Sales<br>+ Informing the Public<br>+ Literary Awards |
| Government | + Public Affairs Dept. | + One-on-one conversation<br>+ Site Inspections<br>+ Leadership Forums | + Getting re-elected<br>+ Compliance with Legislation/Regs. |
| Suppliers | + Purchasing Dept. | + Request for proposal/bids<br>+ Contracts | + Getting the Contract |

*Matrix format develeped by Joel Kurihara and Maria Bober Rasmussen as part of the Global Environmental Management Initiative (GEMI)- 1993*

For ongoing environmental communications, an inexpensive way to proceed is to look for places you can insert your key messages into routine communique: articles in your company newspaper, a mention in your product literature, during your facility open house. If your communication is "event driven" and timing is an issue, you will likely have to use more immediate mechanisms—press conferences, press releases, letters to facility neighbors, door-to-door visits, meetings with employees—to deliver your message. Here, again, you should confer with those in your company who routinely communicate with your key audience. They will know the best route to use to quickly get your message to the right people.

In addition to learning from the other communications professionals within your company, make yourself a teacher. Educate them about your company's environmental efforts and key messages and enlist them to incorporate your issue into their efforts.

## EMPLOYEES: KEY TO SUCCESSFUL ONGOING COMMUNICATIONS

We recently had reason to contact several prominent companies to discuss environmental communication strategies. Over and over again, we heard that employees are the number one key audience. The reasons why are not difficult to understand.

Employees are your first line of communication, and the lifeline of the company. Employees, from the shop floor to the corporate offices, are who make your environmental vision into a reality. They make decisions and take action everyday which define your environmental performance.

Your employees live near your facilities and they, along with their friends and neighbors, make up the community. They are a key influencer of other audiences, and depending upon the information you give them, they can be either your best advocates or your worst detractors.

Keep in mind the advice given at the outset of this chapter: saying nothing is as bad as having nothing to say. And if your employees don't feel good about your environmental efforts, who will?

A framework for employee environmental communications might include two components:

- *Educate*—If your company has an environmental philosophy and specific goals, make sure the employees hear about them on an ongoing basis, so they can act appropriately. The company EH&S policy reproduced on laminated credit card size stock can be sent to all employees as a reminder on what the company expects.

- *Celebrate*—Tell employees about progress achieved on your goals; they will carry those messages to other publics. Another way to celebrate is to feature the good environmental work that employees are doing in your company newspaper or magazine or, if you have the connections, get an article printed in the local newspaper about your employees efforts. Still even more powerful is to recognize employees with a companywide environmental awards program. Some companies do this, and the awards are presented by the CEO or executive management at an annual awards ceremony.

## CRISIS COMMUNICATION

Today news travels fast. Bad news travels even faster. A company needs to be prepared in *advance* to deal with a potential safety, health and environmental crisis situation. This preparation must involve the entire organization, from the CEO down to the facility level, or there could be misalignment in both the physical response to the crisis and the communication response to the crisis.

While the physical response to the crisis is primarily focused on minimizing the immediate and short-term effects, the communications process focuses not only on the immediate informational aspects of the crisis but also on the long-term consequences to the company.

Many chemical and petroleum companies have developed sophisticated crisis response plans which include centralized "emergency response rooms" at facilities and corporate headquarters to coordinate physical response and communications. Environmental, health and safety professionals are key members of this response team as well as personnel from the public affairs organization.

There will be many potential audiences to communicate with in a crisis: the local citizenry where the crisis has occurred, medical and emergency response teams from the community, school administrators and other public officials, regulatory investigators, the local media, peer companies, perhaps the national media, business management, and senior management.

A spokesperson for the company should be assigned to coordinate all communications about the crisis. This individual then, must be given the latest information, corroborated and recorroborated to ensure its accuracy. That spokesperson should have assistance in answering phone requests for information, keeping a log of media and citizen requests, verifying the information that is coming in, preparing talking points, and developing press releases. This spokesperson may decide to utilize others in the organization to deliver particular messages during the emergency situation. Therefore, these selected individuals, such as plant managers, or environmental and safety professionals need advance training so that they will present themselves as knowledgeable, credible company representatives should a situation arise where they may be called upon.

Several things should be kept in mind when communicating during a crisis situation

- Maintain a potential contact list at each facility and at the corporate level;
- Be open with your audience. Withholding information that is later discovered that you had will surely backfire and damage your credibility. In addition, don't appear defensive;
- Never speculate; deal with verified facts. Speculation will lead to disaster, and will most likely lead to a position from which you may have to retreat;
- Provide media access to your facility management and business management. Their message will carry greater importance than that of your EH&S professionals (Sorry about that!);
- Avoid the "no comment" approach. This gives the impression that you are unresponsive, or worse, hiding something that you don't want known;
- Include information for your stockholders, suppliers, and customers, if relevant, in your communications. They, will be interested in overall company impact and in product supply as a result of the crisis.

A crisis situation is not the first time to be testing your crisis management approach. Periodic simulations of emergency situations will help to season your people and expose potential weaknesses in your response plan.

## SUMMARY

The communications process that you use will have a significant impact on the internal and external success of your EH&S efforts. Internally, effective communication can be a driver for your vision and energize your employees to deliver the results necessary for you to achieve your performance objectives. Externally, they can give you the credibility necessary to effectively operate in today's environmentally conscious society, and if you should be unfortunate to suffer an EH&S crisis situation, effective communications can help mitigate the negative impact that such events create.

# 9

## STRATEGIC PLANNING

*Ronald W. Michaud, Pilko & Associates, Inc.*

### HISTORICAL PERSPECTIVE

The complexity of environmental, health and safety (EH&S) issues facing companies today requires a management approach that is very different from that of the early days of the environmental movement in this country. What was once an arena overseen by a few "specialists" and company attorneys has broadened to include the necessary involvement of the line organization, business leaders, R&D and engineering resources, CEO's, and the board of directors of corporations.

There are many reasons for this

- The tremendous growth of federal environmental regulations since the 1970s, their growing complexity and intrusiveness in process and product related issues, and the tightness in timetables for compliance has placed significant additional demands on the organization. Permits take longer to negotiate capital projects require more careful consideration and need to be executed more quickly, and process, product, and packaging changes are sometimes necessary, which requires research and development.

• In addition, public scrutiny is more intense and there is increased public demand for information about a company's environmental impact. Since 1989, companies have had to publicly report their emissions of certain chemicals under SARA 313. Although, in many cases, regulations don't require reductions in these emissions, companies have had to develop proactive emissions reduction efforts to meet public expectations. Making decisions on which streams will be reduced and the timing for achieving these reductions is another decision set added to the regulatory decision set.

• Likewise, worker safety and protection has broadened in focus. The focus of the effort until the eighties was primarily on preventing injuries and acute exposure to toxic materials. To this has been added the dimension of cumulative trauma disorders, employee monitoring for long-term exposure to low levels of toxic materials, communication to employees about potential hazards of exposure on a chemical-specific basis, and process hazards management.

With these multiple issues facing many businesses—issues that will likely increase with time—along with internal efforts to "reinvent the corporation" and improve the cost effectiveness and value of all activities, careful planning is necessary. What was, in the past, a "react to the situation as it arises" management approach for dealing with EH&S issues, now requires a comprehensive strategic planning approach for future success. Companies with vision and goals and who competitively execute their plans using management systems will be able to succeed in a "green" public arena.

## BENEFITS OF STRATEGIC PLANNING

Almost all mid to large companies have a strategic business planning process where they develop three to five year projections (and some even ten year projections) of where they want the business to be at those time intervals. They obtain information on competitors, project cost and sales price for the businesses' products, do market analysis, project capital expenditures for increased capacity, quality improvements, and new product introductions. They also look at their R&D needs and manpower allocations. Until recently, the EH&S aspects of the business were not fully integrated into the business planning process. Some forward thinking companies began to include EH&S strategies into their business planning process in the late 1980s and early 1990s. To bring focused attention to EH&S issues, some companies created a stand alone company EH&S strategic plan that contained input from the business, and which profiled in detail what the company and its businesses were doing in the EH&S arena. After several years, the EH&S issues became fully integrated into each business' strategic plan, and a few companies developed databases which could construct from each business plan a profile of what emissions and waste were being reduced, how much was being spent for compliance, how corporate goals were being met, and what technology was being used. This information was extremely valuable in making decisions on where additional focus was needed, and where marginal improvements were being made at too high a cost.

The strategic planning process, as well as the resultant strategic plan, has many benefits for the organization

- It will create a common vision and align the activities of the various parts of the organization that have major impact on the environmental performance of a company.

- It enables effective deployment and utilization of resources, both in the EH&S function and in other organizations engaged in EH&S activity.

- It will assist in integrating EH&S issues into all business activities so that they become part of the ongoing business decision process.

- It provides the framework for making decisions on discretionary EH&S activities that are prudent from a business standpoint, although they may not currently be required by regulations.

- It will reduce capital and operating costs for EH&S-related activities in the long term, due to better up front planning. It also generally results in the smoothing of capital expenditures over time.

- It provides the ability to capitalize on environmental or safety driven business opportunities.

- It minimizes surprises and the frustration of chasing the "issue of the day."

**The bottom line is that you're more in control of your EH&S efforts with a strategic plan than without one.**

Once a company has made the decision to develop an EH&S strategic plan, it has already taken a step toward a proactive approach to managing EH&S issues. George Pilko, president of Pilko & Associates, Inc., a consulting firm, describes being proactive as "doing what's prudent, from an EH&S standpoint, to support your *long-term business strategy*, regardless of whether these activities are *currently* required by the regulatory agencies." Proactive management, however, is more complex than the reactive approach of meeting regulations as they are promulgated. It involves activities that are more discretionary, using good judgement, perseverance and a focus on the long term. Hence, the need for strategic planning.

## EH&S STRATEGIC PLANNING APPROACH

EH&S strategic planning can be defined simply as "the planned management of change of your EH&S effort over an extended time frame." It is a logical, systematic process that can be broken down into a simplified four-step approach (See Figure 1)

1.  Performing a thorough analysis of where your company stands today in your EH&S effort;

2.  Analyzing those factors or influences that will impact your ability to move from where you are today in your EH&S effort to where you want to be in the future;

3.    Describing what the EH&S effort should look like at some future point in time—typically five or more years.

4.    Developing targets, goals, strategies, and a realistic roadmap and specific plans to get you where you want to be.

Each of these-steps will be discussed more fully in the subsequent sections of this chapter.

There are many other models that can be used for your strategic planning effort. However, they most likely will contain some form of the four-step process described above. You should use a process that fits comfortably within your company culture and that is aligned with the strategic planning efforts of your businesses.

The strategic planning effort is no small task. To do a quality effort that will produce an executable strategic plan will take several months and consume significant manpower from many parts of the organization. However, this front-loaded effort will result in more effective management of your EH&S issues over an extended time frame and lead to improved performance at a potentially lower cost.

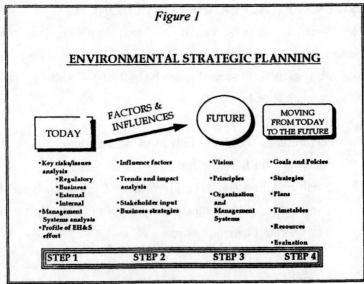

*Figure 1*

**ENVIRONMENTAL STRATEGIC PLANNING**

| TODAY | FACTORS & INFLUENCES | FUTURE | MOVING FROM TODAY TO THE FUTURE |
|---|---|---|---|
| •Key risks/issues analysis<br>  •Regulatory<br>  •Business<br>  •External<br>  •Internal<br>•Management Systems analysis<br>•Profile of EH&S effort | •Influence factors<br>•Trends and impact analysis<br>•Stakeholder input<br>•Business strategies | •Vision<br>•Principles<br>•Organization and Management Systems | •Goals and Policies<br>•Strategies<br>•Plans<br>•Timetables<br>•Resources<br>•Evaluation |
| STEP 1 | STEP 2 | STEP 3 | STEP 4 |

# INTEGRATED PLANNING

To be successful, an EH&S strategic plan needs to be integrated at the corporate, business, and facility levels. The corporate element of the plan will usually address direction-setting components, some of which are described below:

- A company EH&S vision and how that vision will be communicated and reinforced with top management support, using policies, guidelines, and company EH&S goals;
- Major issues that transcend businesses, such as the EH&S positioning of a company opposite its major competitors;
- Establishing relationships and partnerships with external stakeholders, and participation in industry EH&S initiatives;
- Overall EH&S function, personnel development, and deployment;
- Programs and management processes to support the businesses in achieving EH&S goals; and
- The monitoring of emerging EH&S issues and how the company will respond.

With the corporate portion of the plan providing strategic direction, individual businesses manage their EH&S issues by focusing on such elements as

- Product and manufacturing process issues, including pollution prevention and design for the environment;
- Industry EH&S issues;
- The impact of regulations on business profitability, and programs for compliance; and

- The allocation of resources to meet EH&S obligations.

With the business strategic plan as guidance, at the facility level the plan is focused on

- The facility's EH&S impact on its employees, the community, and the surrounding environment;
- Complying with ongoing regulatory change;
- Meeting business needs for expansion or contraction as they relate to EH&S issues; and
- Meeting facility, business, and corporate EH&S goals.

All of these strategic planning efforts—corporate, business, and facility—must be aligned and coordinated so that the resultant strategic plan is consistent in focus, resource allocation, and timing for accomplishment.

In addition to the content of the EH&S strategic plan, careful consideration needs to be given to which organizations and individuals will be involved in the strategic planning process. For an EH&S strategic plan to be accepted and successfully implemented by the organization, those who are most impacted by the plan need to be involved in its development from the beginning. The development of the plan should not be the sole effort of EH&S personnel.

Establishing a *core team* to lead the effort can be an effective means to manage the process. Consideration should be given to having representatives from the following groups on the core team: business planning, operations, engineering, R&D, public relations, EH&S, legal, marketing, and there may be others.

Another aspect to consider during the strategic planning process is the input of various company stakeholders. Valuable information can be obtained by inviting various internal and external stakeholders of varying viewpoints to share their thoughts with members of the core team or an even wider audience. This is particularly useful in Steps 3 and 4, when discussing internal and external trends and influences, and shaping goals. In its major environmental strategic planning effort in 1992, the Dupont Company invited members of academia, the environmental conservation community, environmental policy groups, and customers to share their views on what the developing environmental issues were and how the company might address these issues. Thoughts and ideas were presented that perhaps could not have come from just an internal strategy group.

## Step 1: Where Are You Today?

A careful examination of the present state of your company's EH&S efforts will give the starting point from which to develop future plans. The following areas should be probed at all levels of the company.

- What are the biggest EH&S risks or impact areas that you now face? Are they, for example, in the process safety arena where there is the potential for a catastrophic event, or in groundwater or soil contamination that will require future costly remediation, or perhaps in the risk of penalties due to noncompliance?

- What is the state of your EH&S management systems? Do you operate mostly ad hoc, or are your systems no longer valid since the last company re-engineering? Of the EH&S management systems in place, how many are as effective as you'd like them to be?

- How well-trained and effective are your EH&S and operating personnel in discharging their EH&S responsibilities? Do you have adequate personnel? Do you have a development program to increase their capabilities?

- At what stage are you in your environmental effort? Are you reactive to regulation as it gets promulgated or do you anticipate emerging issues and plan ahead?

- At the business level, what is the EH&S impact of manufacturing processes and products compared with other businesses in the company, or with industry competitors, or that of alternative products?

There are many other questions that could be asked that are specific to your company's issues and needs, but those above will generally apply to all companies. The objective is to get as complete a picture as possible of where you are today with your EH&S efforts and impact.

## Step 2: Analyze Factors and Influences

Analyzing those factors or influences, both internal and external, both positive and negative, that will potentially impact you over the time span of your strategic plan will allow you to adequately address them in your strategies. Some areas to probe are:

- Anticipated federal and state laws and regulations and their impact on company operations. Although predicting regulatory initiatives is not an exact science—there is a good degree of uncertainty in a politically engendered process—a trend analysis can provide useful information. There is little doubt that the number of regulations and their complexity will continue to increase, at least in the foreseeable future. Getting a good handle on their potential impact can lead to early product and manufacturing process changes which will reduce the overall regulatory burden.

- Changes in public expectations, especially in the communities in which you operate;
- Business growth or contraction, including planned future expansions, facility shutdowns, and acquisitions and divestitures. There should also be an analysis of business health—can you afford to be in this business and meet your future EH&S obligations?
- Anticipated product and process changes; and
- Anticipated EH&S technology developments.

The objective in this step is to get as complete a picture of the challenges you will face, both internally and externally, in your EH&S efforts. As mentioned earlier, this step provides the opportunity to seek input from a diverse stakeholder constituency.

## Step 3: Describe the Future

This visionary step can be likened to reading tea leaves. However, I want to stress that the vision must be expressed not as a fuzzy, motherhood statement such as "sustained EH&S leadership" but, in pragmatic and quantifiable terms, it should describe the attributes of your future EH&S effort.

Some examples may be:

- We will achieve and maintain the number one or two position in our industry on the least waste generated per pound of product;
- We will invest, innovate, and collaborate with others in industry, government, and academia to seek new technologies for pollution prevention;
- We will have formal, documented management systems in place for all aspects of EH&S management;
- We will utilize total quality management (TQM) to continuously improve our EH&S effort; and
- We will require sound environmental, health and safety practices from our suppliers and contractors.

It is equally important to describe why you want to be there and how the company and its businesses will benefit from being there. This is an analysis that is often forgotten in the strategic planning process and is the reason why some of these strategic plan implementation efforts are never fully carried out. Too often the vision is vague, without business relevance or thought given to what the benefits would be when the vision was achieved.

This step in the strategic planning process also affords opportunities for stakeholder input and the opportunity to do some benchmarking with other companies to determine what are the best-in-class EH&S practices and performance. Benchmarking of EH&S activities is discussed more fully in Chapter 15.

## Step 4: Develop Strategies and Plans

This step in the strategic planning process consists of a series of sub-steps necessary to achieve your future state.

- Determining a set of policies and goals needed to guide the effort;
- Developing performance measures that are indicative of progress in achieving the future state;
- Defining roles and responsibilities in light of the new future state;
- Establishing management systems to produce predictable outcomes;
- Developing strategies and action plans to move the company to the described future state;
- Establishing resource needs to match the strategies and action plans, and
- Establishing timetables with milestone events

When you get to this step in the process you will begin to see the value of having described your current state as completely as possible, since this is the starting point for the application of strategies and action plans to move you from where you are to where you want to go. Also important, as you develop these strategies and action plans, is to be mindful of the influence factors that you analyzed in Step 2. The strategies and action plans need to take these into account.

## COMMUNICATING THE PLAN

During the development of the EH&S strategic plan, the core team should review its progress with others in various segments of the corporation to (1) do a reality check that the plan elements are reasonable; (2) seek input to refine the elements of the plan and, (3) by doing the aforementioned, begin the process of "buying into the plan." It is important to get top management support for the plan, since they usually control the allocation of resources, as well as from those who will implement portions of the plan, since effective, timely implementation is a prerequisite for success.

Once the strategic plan has been developed and approved, clear communication of the plan to all segments of the organization that will participate in its implementation is necessary.

## AUDITING PROGRESS

Auditing and reporting progress on the plan is a key component of assuring the plan is implemented and successful. It is also utilized to make necessary changes to action plans and strategies if they appear not to be producing the desired results. Also, this is a means of keeping the plan refreshed and evergreen during the course of the year.

Strategic EH&S planning is not a one-time event. Once the vision and some basic strategies have been developed, there begins an annual cycle of review and integration of new action items to move the company toward its vision. This should parallel the business planning cycle and become an integral part of it. Although initially, a separate EH&S strategic plan may be needed to focus effort, it should gravitate to being part of the standard business planning process as soon as practical.

# 10

---

# PROCESS SAFETY

*Ray E. Witter, American Institute of Chemical Engineers (AIChE)*

## INTRODUCTION

In today's society, the public, customers, plant personnel, and regulatory agencies demand that companies take necessary actions to reduce the possibility of episodic hazardous materials incidents.

As the chemical process industries have developed more sophisticated ways to improve process safety, we have seen the introduction of safety management systems to augment process safety engineering. Management systems for chemical process safety are comprehensive sets of policies, procedures, and practices designed to ensure that barriers to major incidents are in place, in use, and effective prevention of chemical process accidents requires the implementation of effective, comprehensive process safety management systems.

## MANAGEMENT OF PROCESS SAFETY

Process safety management differs from the personal safety activities that have been in place in industry for many years. Process safety is the operation of facilities that handle, process, or store hazardous materials in a manner to protect people and property from episodic and catastrophic incidents. Process safety is a dynamic system involving technology, materials, people, and equipment that make up a facility. However, the use of materials with inherent hazardous properties can never be done in the total absence of risk. Thus, process safety is an ideal condition toward which one strives.

Process safety management is the application of management systems to the identification, understanding, and control of process hazards to prevent process, related injuries and incidents. Process safety management concepts should augment rather than be a substitute for the management principles and practices that are in place in an organization.

At every level, the critical ingredient in any management system is leadership. For chemical process safety management, leadership is essential to provide visibility, momentum, a sense of organizational commitment and direction, and ultimately reinforcement, through the distribution of rewards and punishment for variable levels of performance. Leadership is needed at every level, from the CEO to the first line supervisor. In the absence of strong, effective, continuing leadership, the desired level of safety performance will not be achieved.

Chemical process safety requires management systems to provide sound facility design, construction, operation, and maintenance. They assure the establishment of overall process safety goals and the integration of these goals with business and other strategic organizational goals. Safety management system concerns need to be addressed at all organizational levels: strategic, tactical, and task levels.

At the strategic level, process safety management systems are concerned with establishing and reviewing the overall process safety goals and policies of the organization and this would address the need to retain and acquire only those businesses with acceptable risk to the organization.

At the tactical level, the process safety management system would focus on providing information and decision support for assuring that process operations are conducted in a safe manner. A procedure for managing change in facilities or technology is an example of a tactical level system.

At the task level, process safety management systems are directed toward controlling the regular, ongoing activities. These would include writing maintenance procedures, development of standard operating procedures and operator training.

The effectiveness of a process safety management program depends on the assignment of responsibilities, and on holding individuals accountable for their performance. Every element of process safety management requires assignment of accountability and responsibility.

## PROCESS SAFETY INFORMATION AND DOCUMENTATION

The design, construction, and operation of facilities for the production and storage of hazardous materials involve a significant commitment of company resources. A considerable amount of intellectual equity is acquired over the years as a result of this commitment, which is fundamental to the long-term viability and financial success of a company. In addition, this process knowledge is the foundation upon which many aspects of a process safety program are built. Documentation and retrieval of this knowledge are important for process safety for a number of reasons, including:

- preserving a record of design conditions and materials of construction for existing equipment;

- providing recall of the rationale for design decisions during major capital projects;
- offering a baseline for use in evaluating process changes;
- recording accident and incident causes and corrective actions;
- retaining basic research and development information on process chemistry.

In addition, legislation and regulation now require owners of chemical plants to maintain current documentation and procedures on many aspects of plant design and operation.

A properly designed management system will ensure the documentation and retention of safety-related process knowledge. Components of process knowledge and documentation include

- chemical and occupational health hazards;
- process definition and design criteria;
- process and equipment design;
- protective systems;
- operating procedures (normal and upset conditions);
- process risk management decisions;
- company memory.

The process and equipment documentation, as well as operating procedures and incident investigations, should create an archival history of the operation. Also, while it is important to know the current status of the operation, it is important to be able to look back and learn from the unit's history of process safety performance and improvement. Retention of historical information should be done in accordance with the company's record retention policy. There should be specific responsibility assigned for the management of historical records. Issues which should be addressed by the management system include specifying who will keep the records, where and how they will be maintained, security of information, and how they can be retrieved and used at both facility and companywide levels. Consideration should be given to providing backup for critical records and protecting the records against loss.

An important aspect of company memory is the knowledge and experience of senior operators, supervisors, and engineers. A gap in company memory can be created by early retirements and downsizing. Should such situations arise, the process knowledge management system should prompt the initiation of programs to capture as much of this experience-based information as possible. The system might include engaging early retirees temporarily as consultants, or having a younger employee work with the experienced person for a period of time prior to retirement.

## PROCESS HAZARD ANALYSIS

With the public's heightened awareness of chemical hazards and more stringent federal and state government regulation of the chemical and petrochemical industries, project reviews need to be comprehensive and systematic. There is the need for process safety reviews on capital projects, major capacity increases, alterations to existing facilities, new chemistry, new operating procedures, manufacture of new products, and so on..

For capital projects, safety review procedures should interact with all phases of the project. The five phases of a typical major capital project are

Phase I - Conceptual Engineering—involves the technical and economic evaluation of a project's feasibility, including process chemistry, process hazards, instruments and controls, and safety systems.

Phase II - Basic Engineering—involves process calculations, process flow design, piping and instrumentation diagrams, and equipment data sheets.

Phase III - Detail Design—involves vessel thickness calculations, heat exchanger sizing, piping materials and design, equipment specifications, and construction drawings.

Phase IV - Equipment Procurement and Construction—involves purchasing of fabricated equipment, construction supplies, and on-site installation.

Phase V - Pre Startup—involves check out and run-in activities performed to ensure that equipment and piping are mechanically integrated and free of obstructions, determining that instruments and controls are functioning properly, and checking other aspects of the project that should be monitored for safe and effective operation.

Several types of hazard review procedures are available and can be applied in safety reviews. These hazard reviews require different degrees of detailed information. Therefore, the applicability depends on the stage of the project. The hazard review management system should assure that an appropriate review technique is used for each phase of a process life cycle.

Examples of hazard evaluation during the life cycle of a project include:

- Conceptual Engineering - Preliminary Hazard Analysis
Checklist

- Basic Engineering-Relative Ranking
What-if Analysis

- Detail Design-What-if/Checklist
    Hazard and Operability Study
    Failure Modes and Effects Analysis

- Construction-Checklist Safety Inspections

- Startup-Pre-Startup-Safety Review

- Operating Unit - What-if/Checklist
    Hazard and Operability Study
    Failure Modes and Effects Analysis

- Decommissioning - Checklist
    What-if/Checklist

An important component of project safety reviews is the follow-up and closure of all action items. Action items must be addressed and actions documented. This is true whether the initial suggestions were followed, new solutions developed, or no action was necessary. Process safety reviews are a critical element in ensuring that changes or additions to facilities and equipment are done safely.

## Operating Procedures

Most process safety information is used off-line by technical staff personnel in design or modification of the process. Operating procedures represent a special type of process documentation in that they are required by manufacturing personnel for day-to-day operations. Accordingly, the operating procedures should be in place for new and modified facilities prior to start-up.

Operating procedures should provide clear instructions for safely conducting activities involved in each process, and for switching between products in a batch operation. There should be procedures addressing each operating phase, including initial start-up, normal operation, emergency operations and shutdowns, normal shutdown, and start-up following an emergency shutdown or a turnaround. In addition, operating procedures should be developed whenever a temporary or experimental operation is to be conducted. For these and other routine procedures (loading, unloading, preparation for maintenance) a job task analysis may be done and checklists may be developed.

In addition to describing the sequence of actions to be taken in operating the process, the procedures should describe process safety-related operating limits. For each of these limits, the procedure should make clear the consequences of a deviation, steps to be taken to avoid or correct a deviation, and safety systems in place to handle deviations outside acceptable limits.

The operating procedures should also describe safety information related to the process. For example, the hazardous properties of chemicals processed should be described. Where there are special inventory restrictions or quality control procedures associated with the process, these should be described.

The employees who operate the process should have ready access to the operating procedures and they should be familiar with the contents. The procedures should be written in language and terminology clearly understandable to the operators.

It is important that the operating procedures be kept up-to-date. There should be systems in place for ensuring that the applicable process changes are incorporated into the operating procedures and that the procedures are periodically reviewed for accuracy and completeness.

## EQUIPMENT INTEGRITY

Assuring the integrity of process equipment must start with equipment design and continue through its fabrication, installation, and operation. Any equipment used to process hazardous materials should be designed, built, installed, and maintained to control the risk of releases and major incidents. The process integrity program should include

- design standards and specifications;
- fabrication specifications and inspections;
- installation procedures;
- field inspection procedures;
- maintenance procedures;
- training of maintenance personnel.

The criteria for determining what equipment should require a process integrity program should be based on how critical the piece of equipment is to ensuring the safety of the process. Several industry guidelines and government regulations specify the processes in which a mechanical integrity program may apply. For those processes examples of critical equipment are

- vessels and tanks;
- pumps and compressors;
- piping systems;
- relief and vent systems;

- emergency shutdown systems;
- controls, alarms, and interlocks;
- electrical distribution and other utilities;
- fixed fire protection systems.

Process-related maintenance activities should require a written authorization. It is equally important that the inspections, tests, calibrations, and maintenance work on critical equipment be documented to provide a record that can be analyzed as part of the process equipment integrity program.

Process equipment integrity must be considered in all phases of equipment life from its initial design, through fabrication, installation and operation until it is demolished. The program must include quality assurance systems during fabrication, installation, and repair of equipment as well as systems for preventive maintenance during the useful life of the equipment.

## TRAINING

Training is an essential part of any process management system. The proper training of personnel is an absolute requirement for keeping processes operating safely. Training includes making sure that the latest process knowledge is imparted to those responsible for designing, operating, and maintaining the facility. Transmitting the same message to each group tends to reduce the potential for confusion regarding procedures and helps to ensure that consistent actions are taken by all individuals concerned.

As with every other aspect of process safety management, a demonstrated commitment and involvement from the highest levels of management on down is vital to the success of the training function. The well-managed program will have clearly established responsibilities. In addition, those who administer the program, design the lessons, develop course materials and tests, as well as those who instruct, monitor, or audit the program, must be accountable for ensuring the adequacy of performance of all who are trained.

Good training is good communication. It does not just tell the student what he is to do, but how—and especially why—it is important to perform each task, and why it must be performed according to established procedures. Good training may go further by giving the trainee the opportunity to do the required task under controlled conditions or on a simulator.

Components of a training program usually include

- definition of skills and knowledge required (needs analysis);
- initial qualifications assessment;
- selection and development of training program;
- measuring performance and effectiveness;
- qualifications of instructors;
- refresher training;
- records management.

Documentation of training is important for several reasons. It may be required by various regulatory agencies. However, the most important reason is to ensure that proper training and retraining has been achieved as scheduled. Documentation programs generally include

- name of trainee;
- description of courses;
- when the training took place;
- instructor and qualifications;

- what regulatory standards apply;
- method used to verify student understood training;
- schedule for refresher training.

Training is a key element of process safety management. As many other elements of process safety management require specific training to be effective, training need assessment should be conducted for all elements to ensure the completeness of the training program.

## CONTRACTORS

Use of contractor employees requires that these workers are aware of potential hazards in the workplace and comply with the facility's safe work practices. The plant and the contractor need to work together to improve the safety and health of contractor employees.

A facility contractor program may include

- development of a policy that uses specific safety performance criteria for selecting contractors;
- specific safety language in contracts outlining contractors requirements and responsibilities;
- establishment of specific training requirements for contractor employees, including formal safety orientations;
- review the safety and health performance of contractors with the expectation of continuous improvement;
- maintaining copies of the contractors injury and illness statistics;
- auditing the contractors on-site performance on a periodic basis;
- evaluate completed contractor projects for effectiveness of the contractor's safety program.

The contractor's program may include

- understanding of the plant's safety expectations and what is required to improve safety performance;
- development of an accident investigation and prevention program, with emphasis on reducing incidents;
- performing safety inspections on ongoing work;
- training of employees in safe work practices, plant safety requirements, and emergency response;
- documentation of training by names of those trained, the course content, date the course was held, and names of instructors.

Ongoing safety and health discussions between the facility and the contractor are necessary if the contractor safety program is to be effective. Safety issues may arise that might not have been discussed or identified during the initial safety discussions or in the written safety rules. Therefore, the identification and discussion of relevant safety issues should be emphasized throughout the time the contractor is on site.

## SAFE WORK PRACTICES

A critical part of a process management system is the development of safe work practices. Safe work practices normally will address

- confined space entry;
- lockout/tagout;
- hot work permits;
- process line breaking;
- excavation.

Other procedures and permits may be developed to meet the specific requirements of individual facilities, such as lifting over active equipment or piping or high voltage electrical equipment. Other permits or authorizations may be required for testing for flammable or toxic vapors and oxygen concentration. Such procedures should only be valid for a specified time period, typically a work shift. It is important to document work authorizations and safe work permits to ensure compliance with the plant procedures.

Safe work practices are the guidelines for properly and safely performing specific nonroutine operations. Safe work practices and permits are used to

- document responsibility for authorization of work;
- ensure the affected personnel are aware that the work is being done;
- identify the equipment or facility on which the work is being done;
- ensure that appropriate safety precautions are taken prior to beginning the work;
- document that the work has been satisfactorily completed and that all equipment has been returned to an operable condition;
- where necessary, document that the work has not been completed and/or the equipment is still not operable (new permits may have to be prepared and authorized).

Federal and state regulations also require the adoption of safe work practices for certain processes.

# MANAGEMENT OF CHANGE

Changes to processes are made for a variety of reasons, including but not limited to improved efficiency, operability, and safety. Changes can range from large facility expansions or new plants to minor changes in chemicals, technology, equipment, or procedures. Any change represents a deviation from the original design, fabrication, installation, or operation of a process. Even simple changes, if not properly managed, can result in catastrophic consequences.

Three types of changes should be included in the process safety management system at any location

- technology;
- facility;
- organization.

Technical changes include modifications to operating procedures or parameters, use of new chemicals, and revisions to product specifications. Facility changes include modifications that involve substitution of equipment with "not-in-kind" replacements, either temporary or permanent. Organizational changes can range from substitution of personnel to elimination of positions. These changes can have an impact on process safety if they result in insufficient staff or staff having insufficient skills or training, such that they may result in slower or incorrect response to process upsets. Changes in technology or facilities may affect process safety directly, but they can readily be addressed through a management-of-change procedure. Organizational changes have an impact on the accountability and responsibility aspects of process safety.

All such changes need to be identified and reviewed before implementation. At the same time, the process operator must have the flexibility to maintain continuity of operation within established safe operating limits. These safe operating limits need to be made a part of the standard operating procedures. The operator should be allowed to make necessary changes that do not exceed the established safe operating limits. Operation outside of these limits, however, requires formal review and approval by a predetermined procedure. In determining approval level, cost alone should not be used as the criterion.

A challenging aspect of managing change is identifying that the proposed modification is in fact a change. The management of change procedure should provide guidance and examples to assist plant personnel in making this determination. When a change has been identified, the next task is to determine the level of review necessary prior to implementing the change. These important elements must be included in a good management of change procedure.

## INCIDENT INVESTIGATION

"Incidents" can be defined as unplanned events with undesirable consequences. In the context of process safety, incidents include fires, explosions, releases of toxic or hazardous substances, or sudden releases of energy that result in death, injury, adverse human effects or environmental or property damage.

"Near misses" can be defined as unplanned events that could have reasonably resulted in an accident or incident. To some, the definition of "incidents" includes "near misses." Because of the similarities between them, for the purpose of process safety management, "incidents" will be used for both type of events.

Incident investigation is the management process by which the underlying causes of incidents are uncovered, and steps taken to prevent similar incidents. Because the principal purpose of process safety is to prevent incidents, incident investigation is a key element in any effective process management system. Each incident should be investigated to the extent necessary to understand its causes and potential consequences, and to determine how future incidents can be avoided.

Experienced incident investigators know that specific failures point to the immediate cause of an incident, and that underlying each immediate cause is a management system failure, such as faulty design or inadequate training. It is from identifying the underlying causes that the most benefit is gained. This is because by addressing the immediate cause, one only prevents the identical incident from occurring again; by addressing the underlying cause, one prevents numerous other similar incidents from occurring.

In light of the important function of incident investigations in identifying and correcting process safety management system failures, incidents should be looked at as opportunities to improve management systems, rather than as opportunities to assign blame. The cooperation of employees is essential to an effective incident investigation. If the focus of an investigation is to attribute blame, then incidents may not be reported at all, or critical facts may be withheld. The incident investigation should focus on the search for facts in a fair, open, and consistent manner with all employees.

## PROCESS SAFETY AUDITS

Auditing is a critical process safety management element in that it contributes to management control of the other elements. A sound process safety management auditing program will improve the effectiveness of an entire process safety program. In discussing auditing, some confusion over terminology may arise. To assist in understanding the context of process safety management system audits, the following definitions are presented

*Audit*—is a systematic, independent review to verify conformance with established guidelines or standards. It employs a well-defined review process to ensure consistency.

*Process safety management*—is the application of management systems to the identification, understanding, and control of process hazards to prevent process related incidents.

*Process safety management systems auditing*—is the systematic review of PSM systems, used to verify the suitability of these systems and their effective, consistent implementation.

An audit is a fundamental part of an effective process safety management program because its purpose is to verify that systems to manage process safety are in place and functioning effectively. The audit element also needs to have a management system in place to ensure that it functions effectively—particularly the follow-up on action items. Equally important is that auditors have the proper skills and tools to audit effectively.

A comprehensive audit of process safety management systems can be accomplished using different approaches. Audit programs can be developed to meet the needs of a variety of companies from small businesses to international corporations. However, there are basic skills, techniques, and tools that are fundamental to auditing, and some characteristics of good process safety management are common in all facility programs. The information that must be gathered and evaluated during an audit will vary considerably from facility to facility and process to process. Information that an auditor is looking for may reside in more than one location or may not be documented at all.

Regardless of the approach and techniques used to conduct process safety management systems audits, the most important aspects are that the audits are objective, systematic, and done periodically.

## EMERGENCY PLANNING AND PREVENTION

The prevention of incidents relies on good process safety management practices. But even with loss prevention practices in place, incidents may still occur. Whenever hazardous materials are handled, some risk of an incident remains, despite well-engineered equipment, accurate and up-to-date operating procedures, and highly trained personnel. Should a release into the environment occur, a hazard may affect those in the immediate area or adjacent communities.

The time to consider how to handle an incident is before it happens. Good management planning includes identification of the potential scenarios, and preparation of emergency response plans for dealing with a realistic worst case situation as well as minor release cases. These plans may involve dedicated, specialized equipment or facilities. However, in all cases, established procedures for dealing with an incident and treating its effects are necessary.

This emergency response planning embraces a wide range of activities aimed at mitigation or control measures for process upsets, fires, explosions, spills, chemical releases, and other sudden, unplanned events that might result in damage or loss. Emergency response planning objectives might include measures necessary to prevent or limit losses in one or more of the following areas:

- acute health effects to workers, emergency responders, and the public;
- environmental damage;
- property, equipment, or product damage;
- production loss;
- loss of good will and public trust;
- third-party liabilities.

Many aspects of emergency response are covered by federal, state, or local regulations. Before developing and preparing an emergency response plan for a facility, management should determine the applicable regulatory requirements to ensure legal compliance. Examples of U.S. Emergency Response Planning Regulations include:

- OSHA 29CFR1910.3(a)-Employee Emergency Plans and Fire Prevention Plans;
- OSHA29CFR1910.156-Fire Brigades;
- OSHA29CFR1910.120-Hazardous Waste Operations and Emergency Response;
- Resource Conservation and Recovery Act (RCRA);
- Superfund Reauthorization and Amendments (SARA Title III);
- Clean Air Act (CAA);
- National Pollutant Discharge Elimination System (NPDES).

A facility with a well-executed process safety management system will not have the number or size of incidents that would keep emergency responders skilled in the requirements of their assigned duties. Therefore, an emergency plan must be periodically tested by emergency management exercises and drills. Emergency management exercises are essential to assure that training is transformed into an effective response capability. The complexity and scope of exercises appropriate for a facility will depend on the nature of the risks, size of the facility, the legal authority for action by the facility, and the current level of preparedness. A facility with a newly written emergency plan would be ill-advised to test its plan with a full deployment, until that capability has been effectively demonstrated by a series of smaller exercises.

## CONCLUSION

Prevention of chemical process incidents requires the implementation of effective, comprehensive process safety management. Process safety management should be considered as an integrated activity, with appropriate linkages among the various elements, rather than as a series of discrete activities. Through this "holistic" approach, process safety can be most effectively managed.

While an integrated approach to process safety is the objective, designers and managers of process safety management systems should not be intimidated by the potential size of the effort. Process safety management systems are usually approached in a phased manner, with manageable-sized efforts undertaken. Each step in developing the system will contribute to the enhancement of process safety performance.

# 11

## POLLUTION PREVENTION AND WASTE MINIMIZATION

*Jack Weaver, Center for Waste Reduction Technologies,*
*American Institute of Chemical Engineers (AIChE)*

### PREVENTION STRATEGY

Over the past two decades, the impact of complex and demanding governmental regulations has become painfully evident to industry and other regulated bodies, such as municipalities, the U. S. Department of Energy, and the U.S. military. In light of this, the simple concept of preventing waste, emissions, and pollution from being generated in the first place has gained growing acceptance.The benefits are obvious: real cost savings, improved public image, increased competitiveness, regulatory compliance, and the avoidance of major environmental liabilities, such as the Superfund experience. The prevention philosophy, which is a key element of good management practices, is also fundamental to other management systems, such as process safety management (see Chapter 10) and preventive medical care.

# DEFINITIONS

As in many other areas of regulated technology, several conflicting definitions have arisen in the field of pollution prevention. Debating one definition versus another is generally an unproductive exercise. As an alternafive, the following definitions and observations are offered with the suggestion that our efforts will be better spent developing innovative approaches to achieving waste reduction rather than debating the definitions.

## Pollution Prevention

Pollution prevention is generally recognized as a broader goal and includes the prevention of all waste—air, water, or solid—generated from the manufacture, delivery, and use of goods, energy, or services. However, some maintain that pollution prevention includes only source reduction and material substitution.

## Waste Reduction

Waste reduction is regarded by some as equivalent to pollution prevention but by others as a subset or lesser goal. To some, waste refers only to solid or hazardous waste and not to air emissions or wastewater. To others, waste reduction implies end-of-pipe treatment rather than in-process reduction.

## Waste Minimization

Waste minimization is considered to be the ultimate application of waste reduction: pollution prevention. Waste minimization encompasses not only reducing the volume of wastes in all media, but also the toxicity and/or mobility of wastes. Thus, for example, material substitution could result in the same volume of a nonhazardous waste in place of a hazardous one.

## U. S. Environmental Protection Agency (EPA)

The U.S. Environmental Protection Agency (EPA) has adopted a definition which limits polllution prevention to source reduction and materials substitution. It excludes recycling of all kinds (both in-process and post-process) and all end-of-pipe treatment. In communicating with the EPA and others who use this definition, it is important to recognize the differences between these definitions.

## Design for the Environment (DFE)

Design for the Environment (DFE) is a methodology that focuses on the product and process design phases of the product life cycle. Key elements of DFE include the selection of less hazardous raw materials and feedstocks—or material substitution for those already selected—and the development of "green" or environmentally benign chemistry for manufacturing products.

## Industrial Ecology

Industrial ecology applies principles from natural ecology to industrial operations, with a particular emphasis on the efficient use of materials and energy and the elimination of waste. Industrial ecology views industrial systems and infrastructures as if they were a series of interlocking artificial ecosystems, interfacing with the natural global ecosystem.

## Sustainable Development

Sustainable development has been defined in many different ways, but its primary focus is on the efficient utilization of material and energy resources, especially conserving non-renewable resources for potential use by future generations. Sustainable development also recognizes the importance of social equity and the overall impact of humans on the environment.

## REGULATORY BACKGROUND

The primary regulatory framework for pollution prevention and waste minimization includes the major compliance regulations governing industry and other enterprises by limiting the output of emissions and wastes in each of the primary media—water, air and solid. In the United States these include

- Clean Water Act;
- Clean Air Act;
- RCRA;
- Superfund (CERCLA).

With regard to pollution prevention itself, the U. S. Congress passed the Pollution Prevention Act of 1990 which includes the following provisions:

- creates a national policy to reduce or eliminate the generation of waste at the source;
- directs the EPA to undertake a multi-media program of information collection, technology transfer, and financial assistance to the states to promote the use of source reduction techniques;
- requires all facilities reporting SARA Title III (Toxic Release Inventory or TRI) emissions to report toxic chemical source

reduction, recycling, and treatment plans, comparing emissions to previous years as well as to their plans (starting in 1992).

In addition, more than 25 states have laws and regulations addressing pollution prevention, waste minimization or toxic use reduction.These combine measures to build awareness of the need for pollution prevention with a system for reporting both emissions and waste reduction plans.

In 1993, President Clinton announced the Pollution Prevention Executive Order (12856) requiring federal facilities to report toxic chemical releases as part of the TRI, starting in July 1995, to establish voluntary goals to reduce total releases and off-site transfers of toxic materials by 50 percent by 1999, to review and revise procurement procedures, and to develop written strategies for source reduction. The executive order also established a Federal Government Environmental Challenge Program to recognize outstanding environmental performance.

## MANAGEMENT APPROACH

### Pollution Prevention in Commerce

Pollution, emissions, and waste may be generated at every stage in the life cycle of a product. Therefore, pollution prevention, in its broadest sense, needs to consider the total impact of a product or service on the environment from "cradle to grave" or from resource deployment to final product waste management. The stages in a product's life cycle include the following:

- Extraction or deployment of resources, such as mining coal and other minerals, drilling oil and natural gas, generating biomass for feedstocks, etc.

- Manufacture or production of goods—both discrete parts and assemblies as well as continuous process streams and fluids;
- Agricultural production of food and other commercial goods;
- Distribution of manufactured or agricultural goods from the point of manufacture or production to the end user, including transportation, warehousing and receiving, and redistribution services and facilities;
- Use of products for their intended purpose or other purposes;
- Recycle or reuse of products for re-manufacture or for other intended or unplanned purposes (e.g., used tires as fillers for asphalt road surfacing);
- Disposal of used or unused waste products after their useful commercial life—by landfilling, incineration, other treatment, permanent storage, or other means.

## Management Techniques

A number of useful techniques and methodologies have been applied to achieve pollution prevention goals, and many more will certainly be developed. The following techniques are offered as good starting points for any pollution prevention or waste minimization program. The first deals with measuring performance as a guide both for setting reasonable goals and for monitoring progress. The second suggests integrating pollution prevention efforts into an overall TQM program, which might also include other objectives such as process safety and preventive medicine. The third is a step-wise approach for establishing an overall pollution prevention program for existing operations.

## *Pollution Prevention Metrics*

As with any other desired improvement, it is essential to be able to measure performance in order to track progress and establish meaningful goals. A range of different metrics may be utilized, depending, to some extent, on the priorities of the organization or the usefulness of metrics already being used. Some useful measures to consider are

- absolute quantity of waste generated (before any recycle or treatment);
- absolute quantity of waste recycled (both in-process and post-process) relative quantity of waste generated (e.g., per unit of product);
- relative quantity of waste recycled (e.g., percentage of initial waste; or per unit of product);
- concentration of waste (e.g., in wastewater or air);
- reporting data (e.g., TRI and hazardous waste biennial reports);
- number of pollution prevention projects;
- number of employees trained in pollution prevention principles and techniques
- portion or percentage of environmental budgets (capital and expense) allocated to pollution prevention (versus remediation and treatment).

## *Integrating Pollution Prevention into TQM*

Pollution prevention (and related programs such as waste reduction, DFE, life cycle assessment, industrial ecology, etc.) requires a disciplined management approach, as do many other organizational demands. For example, employee health and safety, process safety management, and product quality improvement also require a management systems approach. One of the best ways to accomplish this is through a total quality management (TQM) systems approach, using the principles and techniques of Deming, Cosby, and other leaders in TQM. In fact, pollution prevention fits almost ideally within the TQM framework, as demonstrated by the following parallels:

| TQM | Pollution Prevention |
|---|---|
| Customer satisfaction | Stakeholder satisfaction |
| Continuous improvement | Continuous reduction |
| Management by measurement | Monitoring waste |
| Maximize productivity | Minimize waste |
| Zero defects | Zero emissions |

Before applying the principles and methods of TQM to pollution prevention, it may be instructive to review the basic tenets of TQM.

- a focus on customer satisfaction and achieving a low-cost position;
- the philosophy of continuous improvement;
- techniques and methodologies (management by measurement, use of statistical techniques, systematic workplace processes, and workplace discipline);
- the critical role of human behavior (full stakeholder involvement, need to modify behavior).

## *Policy Deployment*

An effective methodology, borrowed from the TQM arena, for achieving major organizational goals is policy deployment. First, top management of an organization establishes key overall objectives—both longer term strategic and shorter term annual goals. Overall organizational goals or policies may relate to any area of performance, including pollution prevention and waste minimization. Examples might include an overall reduction in TRI releases or a reduction in total waste generated per unit of product. These objectives or "policies" are then delegated or "deployed" throughout the organization. Each organizational unit first identifies the means by which it can best support the overall policies and then proposes specific unit goals. The unit goals are negotiated with management, and progress is reviewed periodically. This has proven to be an effective approach in a number of Fortune 500 companies, resulting in significant reduction in emissions.

## *Establishing an Overall P2 Program for Existing Operations*

To achieve significant progress in pollution prevention (P2) or waste minimization, it is necessary to establish an overall organizational program or initiative and to communicate and monitor it broadly throughout the organization. Such a program might look like the following example:

- conduct an initial waste assessment to identify opportunities to eliminate or reduce waste;
- provide systems such as P2 metrics, to measure and monitor waste;
- analyze the economics of waste reduction such as investment required, savings realizable, costs averted, etc.

- perform limited risk assessments to determine environmental impact of waste streams and urgency for action;
- prioritize waste streams for action and for resource allocation;
- establish specific goals for waste reduction or elimination;
- train staff to implement pollution prevention/waste reduction plan;
- communicate the plan and goals to all employees and other stakeholders.

## TECHNOLOGY APPROACH

### New Product Orientation

Contrary to a common but narrow view, which associates pollution prevention exclusively with manufacturing, the most appropriate starting point for a sound pollution prevention program is at the very start of the commercial process—the initial identification of a product need. The concept of substitution or modification can then be applied not only to the use of hazardous materials in manufacturing, but also to the basic need for the product (or lack thereof) as currently defined or proposed. A sequential approach when dealing with new products could be as follows:

- conduct basic market research to define needs for products and/or services including a critical testing of the needs;
- explore alternative means for satisfying the market needs including a range of products and/or service;
- after selecting a given product/service, explore a range of approaches for—process alternatives—manufacturing the product or providing the service; apply principles of risk

assessment and lifecycle   assessment to select a preferred process;

- apply the principles of waste minimization to reduce the net generation of waste from the product manufacturing and/or service delivery system;

- extend the analysis to include the distribution and delivery of the product and service to the ultimate customers;

- extend the analysis to include the ultimate disposal or recycle of the product after use or depletion by the ultimate customers;

- compare the overall environmental impact (risk) of the selected product and/or services to other alternatives already available or known to be under development;

- implement the selected product/service delivery system.

## Material Substitution

In the case of an existing product with committed manufacturing facilities and other capital infrastructure, it is often difficult to make radical changes in product or process due to economic constraints. However, in this case, a feasible option may be to substitute materials for those presently employed. This could apply to the final product itself or to any of the raw materials or other components utilized in manufacturing the product. For example, in the recent case of chlorofluorocarbons (CFCs), a number of alternative but chemically similar compounds have been developed to yield similar end-product performance with decreased adverse environmental impact. In a number of chemical manufacturing processes, chlorinated solvents have been replaced with less hazardous non-chlorinated solvents, or even with water in a few cases. While some additional capital may be required to modify manufacturing facilities, it is usually much less than would be needed for a totally new plant.

## Pollution Prevention in Product Use and Application

Pollution, waste, and emissions are generated not only during the manufacture of products but also during their intended use and application. In this case, waste reduction may be achieved either (1) by modifying the method of applying or using the product or (2) by modifying the product itself such that emissions and/or waste are reduced during its use or application. Several examples follow:

### *Air Emissions*

Traditional solvent-based paints and coatings, when applied to substrates such as metals or composites, result in vaporization of solvents and other volatile compounds contained in the coating. This has been a common practice in many industries such as automotive, appliances, and aerospace: the VOC emissions related to applying coatings to such substrates can be reduced or eliminated by such techniques as increasing the solids level of the coatings, replacing the solvents with less hazardous solvents, changing from a solvent-based coating to an aqueous coating or to solid powder coatings. Alternatively, in some cases the coating might be eliminated altogether by mechanically or otherwise treating the substrate.

## *Wastewater Contamination*

Water has been used universally to wash products, containers, tools, piping, and many other systems to remove dirt, solvents, food, and other contamination. By modifying cleaning operations to include more dependence on mechanical techniques or the recycling of spent wash water to pretreat or pre-wash contaminated items, it may be possible to reuse the total quantity of water consumed. In some cases, wastewater is contaminated with cleaning agents such as detergents, which are highly stable and active either in natural water receptors or in waste treatment systems. Better selection of detergents, including design of detergents to match the dual needs of cleaning and biodegradation has proven to be an effective means of reducing wastewater contamination.

## *Agricultural Soil Contamination*

A major source of pollution in our overall ecosystem is the contamination caused by the use of agricultural agents and chemicals. This affects first the soil, then both groundwater and surface water and run-off. Thus far, this source of pollution has not been widely addressed, except for moderating the level of usage of agricultural agents. However, it is likely that systems for recycling surface water run-off could be devised to limit the problem, with eventual recovery of the chemicals either for reapplication or off-site destruction or disposal.

## New Products from Wastes

Examples of innovative use of wastes abound and have even been encouraged, in some cases, by governmental action (financial incentives, regulatory allowances, etc.). Scrap plastics have been recycled and reused in a number of different ways. These include the use of segregated, recycled PET or polyethylene (for packaging) or the use of mixed plastics to form composite construction products (traffic barriers, park benches, etc.). Another example is the use of waste rubber tires as an additive in asphalt for highway surfacing.

## MANUFACTURING

In order to implement a pollution prevention program within a manufacturing operation, a range of technical approaches is possible, including the following:

- segregate wastes for recycling, treatment, or toxicity reduction ;
- modify wastes for alternative uses, such as commercial sale or in-plant recycling;
- find substitutes for toxic ingredients;
- improve the utilization of feedstocks and energy by means of process changes;
- utilize separations technology to recover toxic or valuable contaminants.

## Source Reduction

One of the most far reaching means of achieving waste reduction and pollution prevention is by source reduction, that is, eliminating or reducing the generation of wastes and emissions at the source during the initial manufacturing operation. However, source reduction may also be difficult to achieve, due to both technical and economic constraints. Some of the approaches used effectively include:

- *Improved catalysis*—use of novel or improved catalysts to increase reaction yields selectivities, and conversions;
- *Novel reactor design*—use of specially designed reactors to provide more uniform, consistent, and predictable process conditions, and avoidance of local conditions (hot spots, etc.) resulting in unwanted side reactions and lower yields or low selectivities
- *Basic process research*—improved understanding of the process chemistry and mass and heat transfer phenomena, as well as key operating parameters, leading to knowledge of preferred operating regimes to minimize waste generation by maximizing yields and selectivities;
- *Improved process control*—more careful and accurate control of process conditions (temperature, pressure, concentration, etc.) to maintain operations within a preferred operating window for maximum yield and selectivity.

## In-Process Recycle

Regardless of the waste generated at the manufacturing source, reactor or other operating unit, in most processes the overall net process yields can be improved by the "in-process recycling" of process streams. For example, a typical effluent from a chemical reactor may contain the following components: the desired product, some unreacted raw materials, undesired byproducts, carrier solvent, and perhaps some catalysts or other additives. In many manufacturing processes it is advantageous to separate this effluent stream in one of several ways to retrieve unused reactant, solvent, or catalyst for recycle. In some cases, it may be desirable to recycle the unwanted byproduct in order to suppress the side reaction. Many separations technologies are available for accomplishing such a separation. Examples include solvent extraction, steam stripping, solid adsorption, reverse osmosis, ion exchange, distillation, vacuum stripping, and fractional crystallization. Extensive research has been conducted on this wide range of technologies, many of which are directly applicable to in-process recycling. Recent interest has been focused on coupling the design and operation of chemical reactors with that of downstream separations devices called "separative reactors." By co-designing a reactor and a downstream steam or vacuum stripper, for example, the result may be an improved system, the objective is to minimize the net amount of waste exiting the combined system.

## Raw Material Substitution

As noted previously, pollution prevention deals not only with reducing waste volume but also with reducing the toxicity and hazardous impact of wastes and emissions on humans and the environment. In many instances, the environmental impact of a product or process can be lessened by substituting one material for another, without adversely impacting product performance. Examples include the substitution of aqueous solvents for chlorinated hydrocarbons in cleaning fabricated metal surfaces and the replacement of chlorine in water treatment plants by ozone or other oxidizing agents.

## Energy Efficiency

The goals of improved energy efficiency and reduced waste generation are often interlocked and mutually self-supporting. They both stem from the overarching goal of sustainable development, namely to improve the utilization of our total resources—both renewable and nonrenewable and both fuels and raw materials. The two spheres of non-nuclear energy efficiency and pollution prevention are linked in a critical way with regard to fossil fuel utilization, since most fossil fuels can also be used as raw material precursors for producing many organic chemicals. Therefore, the efficient utilization of fossil fuel energy takes on particular significance, since it affects both the generation of wastes and emissions—including greenhouse gases—during the power generation process, and the depletion of nonrenewable resources which also have the potential to be used as raw materials for manufacturing. The benefits of efficiently generating fossil fuel energy and of efficiently utilizing energy are therefore manifold and compelling:

- conservation of nonrenewable energy resources—coal, oil, or natural gas;
- potential to use the fossil fuels as manufacturing feedstock precursors;
- reduced emissions and other waste created in power generation;

- reduced generation of greenhouse gases;
- technical synergy often experienced between energy conservation and waste reduction efforts.

## FUTURE CHALLENGE

In order for our industrial society to sustian a stable and promising future, many major behavioral changes will be required. What will be required is a return to the values and behavior of an earlier era when the primary constraints on our consumption behavior were economic rather than environmental.

As an organization anticipates future requirements for successful performance in the environmental arena, a number of preparatory steps may be taken:

- introduce a comprehensive product stewardship program;
- reduce pollution caused by a customer's use or disposal of your product; that is, looking "downstream" from the manufacturing operation into the user community;
- develop a product substitution strategy to minimize waste;
- use pollution prevention criteria when developing new or improved products;
- perform life cycle assessments and life cycle cost analyses for all new products under development;
- develop and pilot new technologies for waste minimization;
- develop a plan to protect nonrenewable resources by either improving utilization or finding renewable substitutes;
- communicate progress and achievements regularly to all stakeholders—employees, customers, and the public.

Instituting a sound pollution prevention program—based on a good strategic plan, integrated into the overall organizational business plan—makes good business sense. It is the best way to be prepared for future regulatory and business requirements. This proactive approach will require some additional investment now but is sure to yield many rewards and benefits in the not too distant future.

# 12

## DESIGN FOR ENVIRONMENT: MANAGING FOR THE FUTURE

*Braden R. Allenby, AT&T*

### INTRODUCTION

Among the many pressures on enviromental managers in the modern firm is the realization that environmental regulations and the underlying understanding of environmental issues is changing; yet the management of environmental issues for the firm has not evolved in tandem. In many cases, the result is an attempt to manage environmental affairs for the future by relying on past organizational structures. This effort is doomed to failure.

The principal reason is that our understanding of the world has changed. In the fundamental sense discussed in Kuhn's *The Structure of Scientific Revolutions* (1962), we are in the middle of a paradigm shift, and both our intellectual frameworks and the policies and organizations derived from them are no longer adequate. This is not just of academic interest, but has significant management implications as well: it means that simply tweaking existing operations is unlikely to be adequate to manage environmental issues in the future. Rather, we must begin by understanding why the old intellectual framework was inadequate, and then building a replacement to guide us forward. Based on that model, we can then discuss the management strategies appropriate for the future.

## WHY DO WE NEED TO CHANGE?

The simple answer is that we are now beginning to recognize, both as a society and as individual firms, that environment is a strategic issue, not just an overhead cost for our businesses. In the past, environmental impacts had been considered at the macro level to represent simply bad practice curable by remediating whatever local media had been abused. Thus, as shown in Table 1, regulation and legislation generally focused on single media, individual substances, or specific sites, watersheds, or airsheds. Analogously, environmental activity in the firm focused on compliance activities directed at particular streams of emissions from specific facilities, specific substances (e.g., DDT or PCBs), or specific sites where residuals had previously been deposited. In effect, environmental impacts were treated by both society and firms as avoidable results of ordinary economic activity—in other words, as overhead. There was little recognition, even among environmental activists, that the problems being addressed were fundamentally linked with the global scale and technology set of the modem consumer-oriented economy. As a society, and as firms, we were remediating symptoms, not the underlying causes (Allenby, 1993).

That perspective began to change as the fundamental nature of regional and global environmental perturbations began to become more apparent. No amount of spending on programs such as Superfund would ever mitigate fundamental environmental perturbations, such as loss of biodiversity, global climate change, loss of arable soil, inadequate water resources, or world-wide ecological degradation. Similarly, the inadequacy of ad hoc, nonsystemic approaches to the impacts of technological systems became apparent. The most complex environmental issues surrounding automobiles, for example, do not arise because of the manner in which they are manufactured; rather, they arise from the use and maintenance of the product and, more fundamentally, the development and maintenance of necessary infrastructure (petroleum production and distribution, building roads and malls, etc.), and the widely dispersed demographic patterns which have resulted, especially in this country. Even more fundamentally, the role of the automobile as the primary cultural artifact in the developed world potentially imposes significant cultural ideological, and political constraints on environmental mitigation efforts.

This new perspective does not, of course, mean that efforts to reduce near-term localized risks resulting from existing emissions and residuals were inappropriate, especially as a first, naive response to environmental concerns. They are, however, grossly inadequate if the goal is the achievement of a long-term sustainable economy. This is to be expected: initial environmental programs were developed before sustainability had even been proposed as a desirable goal for environmental policy, and were designed to achieve far more limited ends (see Table 1). For society, therefore, the initial reaction to environmental concerns was to treat them as overhead—an unnecessary byproduct of economic activity which could be addressed without any significant change in the underlying industrial and consumption systems. The adoption of sustainability as a primary goal is an indication that environmental concerns are now recognized as strategic—that is, critical to the future welfare of society, and requiring a fundamental integration of environment, economic activity, and technology.

*Table 1*

# EVOLUTION OF ENVIRONMENTAL REGULATION

| Time Focus | Activity Focus | Scale/Scope | Environment In Firm | Firm Focus | Driver | Policy Goal |
|---|---|---|---|---|---|---|
| Past | Remedia-tion | Local in Time & Space; Specific Waste Substances | Overhead; Environ-mental Specialists | Liability | Government Environmental Regulation | Localized Clean Up |
| Present | Compli-ance | Local in Time & Space; Specific Emissions Streams | Overhead; Environ-mental Specialists | Liability | Government Environmental Regulation | Meeting Sub-stance-Specific Numberic Standards |
| Future | Design for Environ-ment; Indus-trial Ecology | Regional & Global Envi-ronmental, Economic &Technolog-ical Systems | Strategic Included in Every De-sign, Plan-ning & Oper-ating Organi-zation & System | Market Ac-cess; Product & Process De-sign; Sustain-able Competi-tive Advan-tage | Customer De-mand; Gov-ernment & World Trade Product Stan-dards & In-centives (e.g., Procurement Regulations | Sustaina-bility |

Not surprisingly, this same evolution is occurring at the firm level. Originally, costs of environmental activities for the firm were treated as overhead, both in an organizational and accounting sense: they were essentially unrelated to the core competencies and activities of the firm. Such an approach meant sometimes significant expenditures for remediation and pollution control equipment, but did not in any way imply a significant operational role for environmental professionals in the firm's activities.

This is, however, changing. Sophisticated customers that drive much commercial procurement, such as governments in the United States, Europe, and Japan, are increasingly demanding environmentally preferable products and services, even if the definition of such creatures is unclear. ISO 14000, a set of environmental requirements now under development by ISO Technical Committee 207, will generate a number of organizational methodological and product requirements.[1] Thus, for example, the quasi-governmental Blue Angel ecolabeling scheme for personal computers in Germany requires, among other things, modular design of entire computer systems; customer-replaceable subassemblies and modules; avoidance of bonding between incompatible materials such as metals and plastics; replacement of many existing flame retardant additives with nonhalogenated substitutes; and post-consumer product take back by manufacturers.

For the firm, the critical recognition must be that such new requirements, focused on products and integrating environmental and technological considerations, cannot possibly be met by traditional environmental management organizations or methods. No amount of "end-of-pipe" expertise can redesign a complex electronic product to be modular, or substitute flame retardants in plastic housings, or support a customer take-back program. Moreover, the costs of failure go not to some overhead account, but directly to the firm's ability to market its products, and the cost structure of each product. For example, failure to initially design a product to be easily disassembled at end-of-life, to establish an efficient logistics system to recover and reprocess the post-consumer product stream, and to anticipate and utilize the most efficient end-of-life reprocessing and recycling technologies will not simply raise a firm's overhead. It will cost a firm's product out of the market—regardless of how efficiently the product can be made initially. Environmental issues change from just a source of liability to a potential source of sustainable competitive advantage—or a source of unforseen competitive disaster if a firm is unable to adapt.

Organizationally, the implications of the new generation of environmental regulation, focused on life-cycle impacts of products rather than just on manufacturing, completely change the role of environmental organizations within the company. Rather than all environmental issues being centered in one specialized organization, environmental expertise must be diffused throughout the firm. It must become part of the accounting process, the business planning process, the research and development process, the product and process design activities, and the marketing process. No longer can the environmental organization claim a firm's environmental issues as their own, or be judged only on how proficiently they perform specific environmental remediation and compliance activities. Rather, they must learn how to integrate their knowledge into the firm's myriad of other activities, to support the development of the firm's strategies, to join in teams rather than to own issues. In many cases, this will require far different competencies than those for which environmental personnel were initially selected: they must become strategic and future-oriented thinkers and team participants, rather than highly specialized engineers with well-defined responsibilitie. Under these circumstances, it is perhaps not surprising that many firms have found their traditional environmental organizations among the most opposed to environmental considerations becoming strategic for the firm. It is, after all, the environmental organizations which will be most wrenchingly changed.

## NEW FRAMEWORK

The first step in understanding how to evolve environmental management systems to meet this new challenge is to create an intellectual framework within which issues can be identified and evaluated. This framework, shown in Figure 1, is anchored by the developing, multidisciplinary field known as industrial ecology.

*Figure 1*

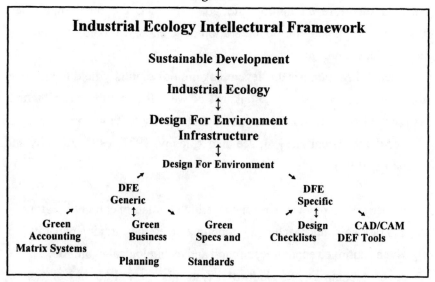

**Industrial Ecology Intellectural Framework**

**Sustainable Development**
↕
**Industrial Ecology**
↕
**Design For Environment
Infrastructure**
↕
Design For Environment

DFE
Generic

DFE
Specific

Green
Accounting
Matrix Systems

Green
Business

Green
Specs and

Design
Checklists

CAD/CAM
DEF Tools

Planning

Standards

*Sustainable development* is both the goal and the vision that supports the framework. Defined as development that meets the needs of the present generation without compromising the ability of future generations to meet theirs (WCED, 1987), the concept is inherently ambiguous. It is also, for some people, ideologically charged, in that in many formulations it implies a need for some form of population stabilization—resisted by many institutions such as the Catholic Church—and a redistribution of wealth from developed to developing countries. It is thus both contentious, and difficult to operationalize. (What does it mean, for example, to design a "sustainable" widget?) Nonetheless, the concept is quite powerful as a goal and, in fact, is being used as such by the Government of The Netherlands, which has the most sophisticated government programs in this area (National Environmental Policy Plan, 1989; National Environmental Policy Plan Plus, 1990; National Environmental Policy Plan 2, 1994). Sustainability in some form is therefore an appropriate goal—in fact, the only appropriate systems-based goal—for the integration of science, technology, economic activity, and environment.

*Industrial ecology* is the developing multidisciplinary field focusing on the system that consists of the linkages between the economic (artifactual) and natural systems. It has been defined in the first textbook on the subject (Graedel and Allenby, 1994; see also Allenby, 1992, 1993; Allenby and Richards, 1994):

> Industrial ecology is the means by which humanity can deliberately and rationally approach and maintain a desirable carrying capacity, given continued economic, cultural and technological evolution. The concept requires that an industrial system be viewed not in isolation from its surrounding systems, but in concert with them. It is a systems view in which one seeks to optimize the total materials cycle from virgin material, to finished material, to component, to product, to obsolete product, and to ultimate disposal. Factors to be optimized include resources, energy and capital. Industrial ecology will form the objective basis upon which choices leading to more sustainable economic activity can be based: it is, to oversimplify, the "science of

sustainability." While it is still nascent, it still offers a number of operational principles, including that which is most fundamental for our purposes: the need to focus on the rapid evolution of environmentally appropriate technological systems if regional and global environmental perturbations are to be mitigated.

*DFE infrastructure* can be thought of as the answer to the question: What does society, and therefore government, need to provide so that properly incented individuals and firms can begin to improve their environmental performance? In general, necessary elements of the DFE infrastructure will include: (1) providing the tools which, for a number of reasons, may not be created by private entities; (2) providing the boundary conditions, including incentives and disincentives, which motivate appropriate behavior on the part of private entities; and, (3) resolving conflicts among regulatory and legal regimes as environmentally appropriate programs are developed and deployed. An example of the first element would be a materials database, which provides firms with information supporting their choice of the most environmentally appropriate material for their needs. An example of the second might be the implemention of an efficient product or packaging takeback system (such programs on a forced, rather than incentive, basis are somewhat controversial and, if improperly structured, can be both economically and environmentally sub-optimal). An example of the third might be reconciling government procurement requirements, which, in an effort to avoid vendor fraud, forbid the purchase of recycled or refurbished products, which are in many cases environmentally preferable.

The *DFE level* is the one where most firms will begin their program evolution. It consists of two components: generic DFE and specific DFE. Generic DFE encompasses those activities which broadly improve the environmental performance of a firm without focusing on specific products or manufacturing processes. Examples might include breaking environmental costs out of overhead accounts in management accounting systems, thus facilitating their active management (Todd, 1994); ensuring that environmental considerations become part of business and strategic planning; reviewing company standards and specifications to ensure that recycled materials are specified as the default option; and developing a "green standard components list." Specific DFE includes the development of "rules and tools" to be applied to the design of specific products and processes. Examples include checklists to evaluate the environmental characteristics of products, processes, and even facilities across their lifecycles (Graedel and Allenby, 1994; Allenby and Graedel 1994); matrix systems to make high-level assessments of individual products or processes (Graedel et al., 1994); or mechanized design tools to apply various possible design choices to obtain environmental figures of merit (Sekutowski, 1994).

## MANAGING THE INTEGRATION OF TECHNOLOGY AND ENVIRONMENT

Managing the integration of environment into a firm's technologies and organizations is a nontrivial and complex task. Before discussing some ways in which this can be done, however, four cautions should be emphasized. First, don't lose compliance. The fact that a number of new demands are being made on environmental personnel, and new skills required of them, is no justification for a firm to let its basic compliance programs slip. DFE programs can, if properly implemented, win over time, reduce the exposure of a firm to regulations, and thus reduce resources required for compliance activities. This will require time, even under the best of circumstances, however, and certainly will not occur in the short term. Compliance programs should not be scavenged to support DFE.

Second, the kinds of organizational changes that are required for the integration of environment into the firm's products, processes, and operations implies a very fundamental cultural change, not only within the firm, but among the firm's suppliers and customers as well. It also requires the development of new tools, new competencies, and new skill sets among numerous personnel. Moreover, in many cases, significant advances at the private firm level will require information or methodologies which do not exist yet, or a corresponding evolution of national policies. Accordingly, only those who misunderstand the nature of the task will believe it can be accomplished easily or quickly. At AT&T, for example, we anticipate that full implementation of DFE will be at least a decade's work, possibly much longer. Do not expect quick fixes, and do not lead your internal customers to expect them.

Third, it must be recognized that DFE is fundamentally not just about technology, but about culture change. Experience in the electronics industry to date indicates that organizational and psychological barriers to the introduction of DFE are frequently far more significant than lack of appropriate technology. It is therefore necessary to initiate all of the activities usually involved in culture change in complex organizations: champions must be identified and supported, reasons for resisting change must be identified and managed, and methods of introducing DFE so that it is least threatening and closely resembles existing practices must be developed. Strong rationales for the program specific to the firm's activities should be developed and disseminated throughout the firm. More subtly, but equally important, the tendency to treat organizational culture issues (which are difficult and complex, and require time to manage) as technological or R&D issues (which are amenable to rational, relatively rapid solution) must be avoided. Initially, in fact, it is probably more important to concern oneself with changing corporate attitudes and thinking, rather than worrying too much about the accuracy of the lifecycle methodologies used. All the fancy DFE methodologies in the world will not save a firm which still considers environment to be overhead, not strategic.

Along these lines, the final caution is that DFE is not, and in most firms never will be, a traditional E&S function. This is partly a matter of tactics: financial personnel will put a lot more credence in a green accounting module that comes from the CFO's office than one coming from E&S. Similarly, engineering and technical managers are more likely to trust and use design tools coming from their R&D organizations and product development laboratories, than those coming from the E&S organization. In these cases, as in others, this is simply a recognition that the E&S group may be competent in environmental issues, but not in financial methodologies or CAD/CAM tools.

There is also a more basic rationale for this caution, however. The traditional E&S function is overhead, not strategic—a very limited function in many ways. When environment becomes strategic, environmental information must be diffused throughout the firm, to be used in all its operations and functions. This is not a situation for central control, although the information and expertise necessary to inform various other functional units may still come from a central location. To some extent, then, DFE takes environment away from the E&S group, although, by making the expertise strategic to the firm, it may enhance their value in the long run. It is this dynamic that accounts for the fact that, in some instances, some of the strongest initial resistance to DFE may well come from the E&S organization. With DFE, the role of E&S is to initiate and support change, then act as a resource. Virtually all DFE programs will be implemented by non-E&S organizations.

Given the above, E&S managers can take a number of concrete steps to begin the implementation of DFE within their firms:

(1) Establish a DFE organization, no matter how small. Experience has demonstrated that trying to support DFE activities as simply an adjunct to existing operations—"additional responsibilities as assigned"—is inadequate, given the magnitude of the necessary culture and organizational changes required for DFE implementation. It might also be worthwhile to center DFE operations in an R&D or product development organization, rather than in the E&S group. In many cases, this seems to facilitate presenting environment as strategic, rather than as overhead.

(2) Establish a system of teams to guide the implementation of DFE, and to facilitate the socialization of the concept within the firm. At AT&T, for example, we have established an AT&T DFE coordinating team with broad representation from business units and support organizations. Under this umbrella team, there is a series of subteams to address specific programs: the Green Accounting Subteam, the DFE Technical Methods Subteam, the Energy Subteam, the Product Takeback Subteam, and so on.

(3) Establish a training program. In most firms, this will consist of a series of programs, from suitcase courses—which can be tailored for relatively informal presentations to staff meetings, concurrent engineering teams, and other small groups—to formal, several-day DFE training programs for managers, engineers, and designers. Some universities, such as the University of Wisconsin, have developed DFE training programs that can be adapted to firm requirements relatively easily.

(4) Establish the capability to answer questions. The integration of technology and environment within the firm will raise many questions and conundrums as it goes forward. Such questions will frequently raise issues of law and regulation (domestic and foreign), product positioning, marketing, product and process design, and environmental science.

Moreover, in many cases the methodologies and data necessary for resolution will be nonexistent or highly uncertain. Accordingly, it is important to establish a capability to rapidly access the most sophisticated resources within the firm to respond to such questions quickly and professionally. At AT&T, for example, we have established a DFE Rapid Response Team, which responds to queries from concurrent engineering teams, manufacturing engineers, product managers, and others, within a week. An additional duty of the team is to distribute their response broadly throughout the company, so that uniform policies can be established and followed.

(5) Establish the capability to review and adapt the best DFE methodologies for the firm's purposes. Many lifecycle assessment methodologies available today, for example, are extremely complex and resource intensive, and are not appropriate for complex manufactured articles such as electronics products. Use of the wrong tool can waste significant amounts of money and time, and provide much useless data but little information. Accordingly, at AT&T we have developed a set of matrix tools that permit us to evaluate the lifecycle impacts of complex items such as facilities, computers, and telephones (Allenby and Graedel 1994; Graedel et al., 1994). Obviously, some detail is lost in such global assessments—but they can be performed in one or two days by a DFE professional, can capture most major lifecycle environmental impacts, and are practical in a real business environment. Over time, we will develop a much more sophisticated and targeted set of tools; for now, let the buyer beware.

(6) Choose targets of opportunity. A few successes, whether they involve product redesign or process improvement, cannot only help educate your internal DFE teams and hone your methodologies, but can be used to help encourage other improvement throughout the firm Be careful not to choose showcase projects that are too complex or problematic.

(7) Establish a DFE outreach or external affairs program. In such a cutting edge area, where advances are being made in widely disparate locations and entities, it is important to establish a network of relationships that allow one to keep up with the field and integrate the latest thinking into firm programs. Relationships with other firms, academic institutions, national laboratories, and governments at all levels (local, state, national, and foreign) should be established as part of this activity. Particular attention should be paid to jurisdictions, such as the Netherlands, Germany, or Sweden, that tend to be leaders in their thinking about product-oriented environmental management systems.

Experience has indicated that individual firms will focus, especially initially, on a subset of DFE activities most appropriate to their culture and circumstances. Inasmuch as this will tend to minimize the dislocations involved in implementing change, this is desirable. Moreover, it reflects the state of the art. At this point, no firm is fully implementing DFE (indeed, given the primitive state of methodologies, data, and government policies, it is questionable whether any firm could, even theoretically, fully implement DFE at this time).

Most fundamentally, perhaps, it is necessary to respect your ignorance as you go about implementing DFE. There is much we don't know, and it is highly probable that we are not yet even asking the right questions. In these circumstances, any management systems and organizations that are set up must be highly flexible, change-oriented, and biased toward continued experimentation and improvement. Similarly, the people you involve in the DFE effort must be comfortable with an ambiguous, ill-defined working environment and mission, and the stresses that inevitably accompany being an agent of fundamental change within the corporation.

## CONCLUSION

Government regulation, customer demand, and an increasingly environmentally constrained world are combining to make environment a strategic, not overhead, issue for the modem corporation. This transition can be initiated, and the interests of the firm protected, during the transition period, through the introduction of DFE practices. This raises, however, some significant management challenges which must be met if DFE is to be successfully implanted in the firm.

## ENDNOTES

[1]     It is frequently claimed that ISO standards are voluntary. While this is technically true, ISO standards become, in fact, requirements for firms engaging in global commerce: they are widely adopted in bidding processes around the world, and trickle down to even small firms through supplier contracts.

## REFERENCES

Allenby, B. R., 1992. *Design for Environment: Implementing Industrial Ecology.* Ph.D. dissertation, Rutgers University.

Allenby, B. R. 1993.

Allenby, B. R. and Graedel, T. S., 1994. Facility paper.

Allenby, B. R. and Richards, D. J., eds. 1994. *The Greening of Industrial Ecosystems* (National Academy Press, Washington, D.C.).

Graedel, T. S. and Allenby, B. R. 1994. *Industrial Ecology* (Prentice-Hall Englewood Cliffs, NJ).

Kuhn, T. S., 1962. *The Structure of Scientific Revolutions* (University of Chicago Press, Chicago).

*National Environmental Policy Plan.* 1989. Ministry of Housing, Physical Planning and Environment, The Netherlands.

*National Environmental Policy Plan Plus.* 1990. Ministry of Housing, Physical Planning and Environment, The Netherlands.

*National Environmental Policy Plan 2.* 1994. Ministry of Housing, Physical Planning and Environment, The Netherlands.

Sekutowski, J. C., 1994. "Greening the Telephone: A Case Study," in *The Greening of Industrial Ecosystems*, ed. by B. R. Allenby and D. J. Richards (National Academy Press: Washington, D.C.) pp. 171-177.

Todd, R., 1994. "Zero-Loss Environmental Accounting Systems," in *The Greening of Industrial Ecosystems*, ed. by B. R. Allenby and D. J. Richards (National Academy Press: Washington, D.C.), pp. 191-200.

WCED (World Commission on Environment and Development, also known as the Brundtland Commission). 1987. *Our Common Future* (Oxford University Press: Oxford, Great Britain).

# 13

## Buying And Selling:
## Site Assessments, Acquisition And
## Divestiture Concerns

*J. Richard Pooler, The Darien Group, Ltd.*

*Robert B. McKinstry, Jr., Ballard Spahr Andrews & Ingersoll*

### INTRODUCTION

The last decade has witnessed a sea change in the role of environmental concerns in business transactions. In the late 1970s and early 1980s, all too often, environmental issues were simply ignored or dealt with using a contractual "compliance with laws" representation and warranty. By the late 1980s and early 1990s, the pendulum had swung to the other extreme. In some cases, the specter of an "environmental problem," regardless of its nature, could interminably delay, or even kill, a financial transaction. In other cases, buyers and sellers alike might decide not to enter the market for certain businesses or properties because of the risk or likelihood of uncovering an environmental problem.

The more sophisticated business person, however, should treat environmental risks for what they are—a business risk that, although potentially serious, can be dealt with like many other risks, such as those posed by changing market conditions or anti-trust and tax laws. Pertinent environmental, health, and safety concerns should be identified early and appropriate and relevant investigation of these issues should be incorporated into the due diligence process, at all stages of transactions, to provide the parties with sufficient reliable information to develop the mechanisms to deal with those concerns realistically.

Appropriate due diligence is critical to assure that the parties make correct assumptions when deciding to pursue a transaction. But more importantly, due diligence can allow the parties to take actions that will minimize the risks to both sides and, where risks cannot be eliminated or avoided, to agree to a realistic allocation of those risks.

## THERE IS NO FORMULA FOR ADDRESSING ENVIRONMENTAL CONCERNS

There is no set formula for addressing environmental, health and safety concerns in business transactions. There is not a single legal method appropriate for addressing all environmental risks. There is also no cookbook formula for an environmental investigation, that could be used in any given transaction.

Both one's legal approach and the type of environmental investigation vary according to the nature of the transaction (*i.e.*, is this a business assets sale or acquisition, a stock sale or acquisition, a hostile takeover, a merger, a land sale or purchase, a loan, etc.); the party's goals (*e.g.*, buyers and sellers goals may vary dramatically); the transactional parameters (time, budget, manpower, and accessibility of information); and the nature of the business at issue. Although there is no specific formula for addressing environmental issues in transactions, in all transactions it is critical to do the homework, identify the goals and the relevant issues, identify the risks, collect necessary information, and apply intelligence to the assessment and allocation of risks.

Of course, in approaching these issues, one must recognize the fact that most large organizations, including many lending institutions, must operate through standard operating procedures, which attempt to reduce situations to particular formulae. Although standard operating procedures can be devised to give due consideration to individual circumstances and to provide flexibility to those applying those procedures,[1] it is an unfortunate fact of life that not all procedures are so well-designed. Thus, in a given transaction, the nature of the issues that must be addressed and the investigation that may have to be undertaken may ultimately be guided by imperfect standard operating procedures rather than issues truly relevant to the transaction.

## TYPES OF ENVIRONMENTAL CONCERNS

In addressing the environmental concerns of any transaction, one should start by first identifying the types of environmental concerns likely to be relevant to that transaction. The types of environmental concerns that may affect any given transaction fall generally into three categories: (1) concerns relating to liability, (2) concerns relating to asset transferability, and (3) concerns relating to either operability or developability (depending upon the nature of the property that is the subject of the transactions). Although the last category of concerns may often present far more significant risks in many transactions, due to the many cases of unexpected liability arising for cleanup under the Comprehensive Environmental Response Compensation and Liability Act or ("CERCLA" or "Superfund"), public awareness has focused most heavily on liability risks.

## Liability

There are three distinct types of liability risks, liability for remediation of environmental contamination, liability for noncompliance with environmental laws and tort liability. Liability to remediate contamination may arise under CERCLA or similar state "little Superfund laws," state or federal solid waste laws such as Resource Conservation and Recovery Act ("RCRA"), or state water and groundwater protection laws. The liability to remediate contamination is often unrelated to the question of whether there has been compliance with the law. Liability to remediate contamination can arise from the mere fact that an acquired parcel of property is contaminated or that an acquired business arranged for disposal of a hazardous substance at an off-site location, or owned or operated a contaminated property in the past. Liability for contamination is largely responsible for concern over property transactions. In recent years, both because it can arise without "fault" and because cost of remediation can exceed the value of the assets being acquired.

Other types of liability, however, can be equally serious. One can "acquire" liabilities for violations of the law by buying a company that has violated the law in the past or by acquiring a facility which continues noncompliant operations after the purchase. Environmental laws typically impose substantial criminal and civil penalties—frequently up to $25,000 per day or higher. Traditional tort liability is also a concern that must be assessed when addressing environmental risks in a business transaction.

## Transferability

The second general class of possible environmental concerns that should be considered in any business transaction is the transferability of the asset that is the subject of the transaction. In most cases, this concern does not relate to whether the asset *can* be transferred as much as *when* the asset can be transferred and under what conditions. Transferability is therefore, an issue most related to timing of the transaction.

The issue of transferability most frequently arises where there are environmental permits to be transferred, such as wastewater discharge permits, solid waste permits, air pollution permits, or water supply permits. The need to transfer permits will usually arise if the asset requiring the permit (such as the physical plant) is sold and may even arise in the cases of mergers and stock sales, depending upon the law of the state where the plant is located and the type of permit. Although, in many cases, the permit transfer will occur as a matter of due course, the transfer may take some time and often must be completed before the transfer of "ownership." In some states, for some permits, certain environmental laws may even bar a transfer of a permit to certain transferees who have a history of past noncompliance with environmental laws or who have outstanding violations at other facilities.

In New Jersey, the Industrial Site Recovery Act ("ISRA" formerly known as the Environmental Cleanup Recovery Act or "ECRA") requires certain state approvals in connection with the transfer of a controlling interest over any "industrial establishment," including transfers effected through stock sales and mergers. Before the transfer, contamination on an industrial establishment in New Jersey must be investigated or cleaned up with state approval or the transferor must enter into an administrative consent order with the state guaranteeing investigation and cleanup. Although at one point, ISRA/ECRA-type laws were under consideration in a number of states, the only state which has adopted a similar law is Connecticut. In large part, other states decided not to follow suit because, by the time they were considering these laws, fear of CERCLA liability had resulted in market forces "requiring" equivalent private investigation and cleanup. Many other states, therefore, limit their involvement to the requirement of deed, or other notifications concerning known environmental contamination in transfers of real property.[2]

## Developability and Operability

The third type of environmental concern in environmental transactions falls under the rubric of either (1) "operability," where an operating concern is being acquired or transferred or (2) "developability," where property is being acquired for development. Both concerns relate to the value of the property being acquired. More specifically, these concerns relate to whether the asset being acquired can be put to its intended use and, if so, when.

The issues that should be considered in addressing whether a given property is developable, of course, depend both upon the character of the property and its intended use. Certain features such as wetlands or flood plains, and certain environmentally sensitive features such as threatened or endangered species habitats will affect most types of development and may preclude all development altogether. Other features may affect only certain types of development or use. For example, a stream with very high water quality will restrict or prevent development that will require a discharge of pollutants to that stream. Similarly, the proximity of national parks, of certain national wildlife refuges, or existing exceptional air quality may constrain certain types of development requiring high levels of air emissions.

Operability is a concern where one is interested in acquiring—by purchase, merger, or otherwise—an operating business. When one acquires an operating business, the buyer needs assurances that it can continue to operate the business lawfully and without incurring liabilities. Even where a seller, for example, agrees to keep the liability from past operations and to indemnify for future violations, the buyer still needs assurance that it will be able to continue to operate the business and generate revenues. The buyer, therefore, needs to consider whether the business has all of the necessary operating permits and whether the business can operate in compliance with those permits. It would do a buyer little good if the seller agreed to pick up past liabilities but the buyer could not operate the business for the substantial time necessary to submit a permit application, build pollution control apparatus, and bring a facility into compliant operation.

While it would be exceedingly unlikely for an operating entity to be forced to shut down while bringing itself into compliance, the costs of bringing a noncompliant facility into compliance with environmental laws can be so significant that these costs should be considered when evaluating price and value. Obviously, where a buyer will need to invest capital into pollution control equipment, a company would be less valuable. Bringing a noncompliant company into compliance with environmental laws may also increase the costs of production or decrease the rate of output.[3]

These operability concerns will likely become more significant, particularly as many companies attempt to incorporate the management standards developed under ISO 14000 into their operations.[4] These standards, still under development, will be important for company marketing strategies and may be incorporated into international legal requirements. Incompatible management structures and the attitudes of personnel to environmental compliance may ultimately be more difficult to "fix" than missing equipment or permits. A purchased company, with a set of management objectives and attitudes dramatically different from the buyer, may never be successfully integrated into the buyer's "culture" without personnel reorganization and changes that could so alter the purchased company's operations as to make it unprofitable.

## STRATEGIES FOR EVALUATING AND MANAGING ENVIRONMENTAL TRANSACTIONS AND RISKS

The foregoing environmental transactional concerns can be addressed by a variety of overlapping strategies. These strategies involve techniques for acquiring necessary information and contractual strategies, both for aiding the acquisition of information and for reducing and allocating risks.

### Due Diligence Techniques

In any transaction, some due diligence in the form of records review and/or site investigation is required and possible. Even, for example, in the case of a potential hostile takeover, where the fact that a transaction is contemplated must be kept confidential, there are public sources of information that can be reviewed to identify types of business operations and publicly known liabilities. This information may be relevant to the price the buyer is willing to pay. In a more common situation, one may review both company records and public records and inspect the company's past and present operating sites in order to identify liabilities, existing permits that may need to be transferred, permits that may be missing, and the status of compliance. Through these investigations, the potential purchaser can obtain the information itself, both to assess the value of the company and to allow the buyer to develop a strategy to allocate risks.

A second "due diligence" technique that produces information, and, to a lesser extent, may help shift some risks to third parties, is to request formal legal or engineering opinions on compliance with laws and operability (as opposed to a legal or engineering analysis). In many cases, obtaining these final opinions, rather than informal analysis, will increase costs substantially without a commensurate gain. However, requesting a formal opinion requires an engineer or attorney to focus more closely upon what information is needed and what legal issues must be addressed.

The parties also may take a more systematic approach to the type of information that must be obtained to determine whether a facility is operating legally and can operate in compliance with the laws. To a lesser extent, the technique of requesting a formal opinion may help to shift "risks" of missing information from the acquirer or lender to the attorney or engineer rendering the formal opinion or, more properly, the malpractice carrier of the engineer or attorney. The latter advantage, however, is probably limited since the formal opinion usually specifically relies upon information provided by the parties to the transaction and includes so many caveats that any malpractice action would be problematic. If the formal opinion mechanism is properly used, the identification of areas where caveats and conditions are necessary may help to focus the parties' attention upon risks which should be considered in the context of the transaction.

## Contractual Strategies

There are a variety of contractual strategies for dealing with environmental risks. A contract can include environmental representations and warranties. Representations and warranties are, first and foremost, an opportunity to gather information. Exceptions to use of the representations and warranties occur where there are risks that must be allocated. Requesting a representation shifts the risk of missing or incorrect information to the party providing the representation and warranty.

The second contractual mechanism for dealing with environmental risks is covenants. Most contracts addressing environmental issues include covenants authorizing environmental investigations and providing "outs" in cases where the investigation uncovers material risks not satisfactorily addressed by the contract. While these "outs" typically allow a party to withdraw from a transaction, this mechanism usually simply allows renegotiation of the allocation of that risk or other terms, such as price. A contractual investigation provision that would be more beneficial to the seller would spell out consequences of unfavorable investigation results and provide for adjustments up front, rather than allow for a blanket renegotiation.

It is critical that a covenant address confidentiality. Environmental investigations often require that trade secrets be revealed. More importantly, critical information relating to environmental issues is often contained in audit reports and correspondence for which the parties may wish to preserve a privilege. In order to preserve the confidentiality of this material to the extent feasible, covenants requiring confidentiality and the return or destruction of copies of sensitive or confidential documents are usually critical. It is also critical that these requirements be made applicable to consultants and other parties receiving the information.

Responsibility for communications with government agencies should also be spelled out, particularly reporting requirements. Typically, if a violation of law or contamination is discovered, the reporting obligation, if any, will lie with the owner or operator, not a consultant. Unfortunately, sometimes a consultant creates a problem by reporting an unreportable event on the basis of the mistaken belief that the consultant has an obligation to report. This can be avoided by some early planning and the clear assignment of responsibility.

Covenants may also provide for time to negotiate permit transfers or ISRA requirements. They should assign responsibility for performing these and other activities necessary to consummate the transaction.

Finally, perhaps most importantly, the contract may include indemnification responsibilities. A warranty is, of course, a type of indemnification. The environmental provisions of contracts typically include more specific indemnifications. These indemnifications typically address liability but can also address lost profits that might arise from problems with transferability or operability. Indemnifications come in many varieties.

The indemnification strategies to be used for environmental issues differ little from those used in addressing other types of issues, with one exception. There is mixed and conflicting case law under CERCLA as to how specific a release or indemnification must be. Some indemnifications or releases have been held not to apply to CERCLA liability. However, the cases agree that if a release or indemnification specifically applies to contamination or CERCLA-like liability, it will be as affective as between the parties to the contract.

## TACTICAL APPROACHES FOR THE DUE DILIGENCE PROCESS

How does one collect the information necessary to evaluate environmental, health and safety issues and formulate an acquisition/divestiture strategy? Firstly, we must dispel the notion that there is a "standard investigation" to perform when acquiring or divesting a property or operation. For shrewd businesspeople, there is no such thing as a "standard investigation." The investigation will vary from transaction to transaction, and depends on the type of transaction, type of property/facility, time constraints, and budget constraints. In some cases, one may not want to conduct an investigation at all. However, when one does wish to conduct an investigation, even a minimally sophisticated transaction, involving multiple sites or industrial operations, will usually require a customized aggregation of components from the following three basic types of investigations:

- Contamination Investigations;
- Compliance Assessments;
- Management Assessments.

In the ensuing sections, each of these investigations will be discussed with respect to three key parameters: (1) Why do you do it, why is this type of investigation important, and what types of liabilities are able to be evaluated through the use of that particular investigation? (2) How do you collect the information in the investigation? (3) What are the typical skills and background of persons who perform the investigation? These parameters should assist managers in selecting the proper investigatory tools for a given transaction.

## THE CONTAMINATION INVESTIGATION

### Why You Do It?

CERCLA, RCRA, and many state little Superfund, solid waste, and water pollution laws all may impose liability to remediate contamination on current owners and operators, prior owners and operators, transporters, and generators. Liability under CERCLA and many of these other laws is strict, joint and several, and is not dependent on amount of material, causation, or legality of original conduct. Moreover, liability can be significant. For example, an average CERCLA remediation will cost $30 million. Thus, liability for contamination can be material in many transactions.

Because these liabilities can be material, in many transactions it is important to determine whether the asset being acquired may be contaminated. For example, information regarding contamination would be important in virtually any acquisition of real property, where the liability could potentially exceed the value of the property being acquired. Even where stock is being acquired, an investigation into potential liability for contamination is desirable to determine whether the value may be impaired by future remedial investigations. Similarly, where real property is being taken as collateral for a loan, an investigation into the presence of contamination is important for the purpose of determining the value of the collateral.

There are also situations, however, where an investigation for the existence of unknown contamination may not be desirable. In many old industrial sites in urbanized areas, unknown contamination is likely to exist and is unlikely to have any significant impact on health or the environment, but, if discovered and reported, may require remediation. Where a lender already has an outstanding loan to an industrial borrower who owns such a property and renegotiates the loan's terms without releasing new funds, it would often be imprudent to require such an investigation, unless and until the lender decided to commence a foreclosure action. If previously unknown contamination is discovered in an investigation, remedial obligations may be accelerated and the borrower's ability to repay the loan impaired. Many sellers of older industrial properties may be reluctant to allow an investigation for previously unknown contamination without some commitments from the potential purchases. Such a seller could be put in a significantly worse situation if an investigation discovers unknown contamination triggering a reporting obligation, and the buyer simply backs out of the sale.

Where it is desirable, a "Phase I/Phase II" investigation for contamination is important to allow the purchaser or lender to determine the probable value of a property, *not* to establish the "innocent purchaser" defense to future CERCLA liability relative to the site. Although the inquiry necessary to assess the value of the asset being purchased will qualify the purchases of real property for the defense if no contamination is found, the availability or unavailability of that defense is often irrelevant as a practical matter because the "innocent purchaser" defense is unavailable to any party who acquires title to a property after contamination is known. Thus, even if the defense is available to a purchaser, if the property is contaminated and the contamination is discovered later, the defense is unavailable to subsequent purchasers (including foreclosing lenders) and the value of the property is impaired.[5]

Regardless of the foregoing, the type of investigation required to establish the "innocent purchaser" defense is, in many cases, identical to the type of investigation desirable for economic reasons. In order to establish the defense, a purchaser must perform "all appropriate inquiry" into the current and historical use of the property. That appropriate inquiry standard is to tied to what is deemed commercially reasonable. Thus, in scoping any contamination investigation, the focus should be upon what is commercially reasonable in light of the parties needs and desires, the structure of the transaction, and cost and timing constraints, not upon what is needed to qualify for the defense.

The Phase I investigation is now uniformly the commercially, technically, and legally accepted starting point for an "appropriate inquiry". Phase I is a nonintrusive exercise designed to determine if contamination may be present on the property. The focus of a Phase I environmental assessment is to provide information which will help determine: (1) if the land or facility can be used for the intended purpose; and (2) if the property possesses serious past or current environmental problems. If the Phase I process concludes that contamination may be present, a Phase II investigation may be performed to verify the presence of contamination, more fully characterize the contamination, and estimate the remedial costs to remove the contamination. In some cases, sufficient information is available to indicate that the "Phase II" investigation should be the first phase.

## Collecting Information

A Phase I environmental assessment is the primary type of audit implemented in conjunction with a due diligence effort for noncomplex properties. This assessment is appropriate for property transfers of raw land, commercial properties, or light industrial properties. ASTM has published a standard approach for performing a Phase I environmental assessment. Several standard components are commonly included in the work scope for these projects. These include

- Initial planning meeting;
- Site assessment guideline;
- Public record review;
- Site inspection;
- Company records review;
- Interviews;
- Reporting.

The major steps normally taken to collect required data and determine the environmental status of a property are as follows:

*Initial planning meeting*—A planning meeting should be held before initiating the project to assure that all parties have a common understanding of both the process and work product, and to discuss the following key issues:

- Scope;
- Schedule;
- Confidentiality measures;
- Procedures for reporting problems;
- Safety and security requirements;
- Site-specific issues.

The importance of this step cannot be overstated. The old adage "Proper planning prevents poor performance," applies here, especially for multi-site investigations. Thoughtful discussion of the above issues among the client/attorney/consultant project team can prevent wasted effort, hone the approach, and assure an effective investigation.

*Site assessment guideline*—An assessment guideline should be sent to the facility in advance to help site management prepare for the facility visit. This guideline can be used by the assessment team to collect site information. Information requested prior to the site visit includes chemicals and processes used at the facility, permits, program documentation, training records, monitoring reports, agency communications files, notices of violation, and other pertinent EHS data. EHS information so provided can be reviewed to identify issues of concern, and focus subsequent assessment activities.

*Public record review*—Appropriate governmental agencies should be contacted and file reviews and/or interviews should be performed. Database lists of federal and state information should be reviewed to identify facilities that might impact the site. These include the federal CERCLIS database of all sites with possible contamination issues, and similar state databases. Local record sources may include the following departments:

planning/zoning, environmental health, sewer/water, water resources, engineering, and fire.

*Site inspection*—A site inspection should be performed to visually confirm areas of concern. All areas of the facility should be inspected. Areas not inspected, for whatever reason, should be noted in the report. A staff person with current and historical knowledge of the facility should accompany the inspection team. The inspection should be scheduled so that the full spectrum of site operations can be observed.

*Company records review*—Available environmental records should be reviewed on-site to determine the existence of major required program components, and identify potential risk issues. An additional strategic purpose of the records review is to perform a cross comparison of site operations and activities against the records, making sure that records accurately reflect operations and activities performed on site and, conversely, that site operations and activities are accurately depicted in the records.

*Interviews*—Interviews are the most critical variable in the entire evaluation process. Good interviewing skills can reveal nuggets of information not available elsewhere. Site personnel should be interviewed to define past and current operational activities relating to hazardous substance usage, hazardous waste generation, and waste disposal practices. Key individuals include those with specific knowledge of site operations, as well as those who have worked on-site for several years. While computer and electronic recording of interviews may be administratively useful, most interviewees will be uncomfortable with devices that document each and every word. This negative impact on the interview process usually outweighs the utility such devices provide.

*Reporting*—The parties should agree upon a report format beforehand. In all cases, recommendations and conclusions should be set forth for all issues. If a definitive recommendation and conclusion cannot be reached on a given issue, additional follow-up actions to further define the issue should be set forth. If the project schedule demands continuous feedback of results, a preliminary verbal report can be performed at the conclusion of the site visit.

In situations where there is clear evidence of contamination, or a high probability of contamination, any recommendation to perform a Phase II sampling investigation must be quickly communicated to the purchaser. Potential and uncertain remediation costs must be considered as early in the transaction as possible. The decision on whether to undertake the Phase II investigation should be made only after carefully weighing the alternatives and the parties' needs. Physical investigations will not fully define likely remediation costs, they will only help to reduce the range of possible costs. As one conducts more investigation, one will generally reduce the range but will spend more time and money. Moreover, because a Phase II will generate new information, it may trigger reporting requirements and accelerate cleanup obligations. Contractual alternatives to allow risk sharing are therefore often preferable to the risks, costs, and delays incident to conducting a Phase II.

Many different strategies can be applied in assuming responsibilities for both investigation costs and likely remediation costs and liabilities. One party can assume all costs, the parties can allocate based on a predetermined percentage, or deductibles can be negotiated (*e.g.*, seller pays first $500,000, with both parties sharing the remainder). Specific remedial tasks may be assigned and unknown and known, contamination problems may be allocated differently. Many options to fairly allocate cost and risk are possible. Regardless, it is critical that in all instances, both parties must agree on the Phase II strategy; if the transaction is completed, the purchaser will be affected by follow-up remedial activities, while if the transaction collapses, the seller will be required to follow through on his own.

## Who Can Do It?

If trained properly, an experienced junior technical expert (engineer, scientist) can effectively perform these assignments. However, appropriate training is key! Phase I environmental assessments are *risk based* evaluations. Many technical personnel have some experience with regulatory compliance issues and will sometimes inappropriately focus on regulatory minutiae. Such detail is rarely significant to the overall transaction. Senior personnel overseeing Phase I environmental assessments must assure that risk is the primary focus of the investigation, and that only significant compliance issues are raised.

## Other Types of Contamination Investigations

There are instances in which other types of "contamination" investigations may be appropriate. For example, in the case of a potential hostile takeover or unannounced tender offer, confidentiality may preclude interviews, site visits, agency record reviews, or any other activity that might suggest that a transaction is contemplated. Nevertheless, a great deal of information can be obtained from electronic research services, which will provide some order-of-magnitude information concerning the scope of likely liabilities.

# THE COMPLIANCE ASSESSMENT

## Why You Do It?

A compliance assessment is used to determine the general compliance of individual environmental, health, and safety (EHS) regulatory programs. In an acquisition/divestiture process, this investigation serves to provide the purchaser with an evaluation of compliance conditions and the likely costs to achieve compliance. The intent is to assure that the major compliance requirements of each regulatory program are being met so that the facility can continue to operate in compliance, and assure that past compliance issues do not create an undue impact (either tort or penalty) on operations.

## Collecting Information

Compliance assessments are detailed processes and operation-oriented regulatory reviews which evaluate a facility's adherence to specific regulatory requirements. The process is divided into three main segments, which include the following tasks:

### pre-audit activities

- Project opening meeting;
- Pre-visit questionnaire to facility;
- Review of relevant regulations and preparation of audit protocols;
- Definition of audit scope and team responsibilities.

## on-site activities

- Audit opening meeting;
- Records review;
- Staff interviews;
- Facility inspection;
- Close-out meeting.

## post-audit activities

- Debriefing;
- Audit report.

The results of compliance assessments are typically presented as findings of noncompliance, which identify the specific regulatory requirement and provide a citation to that requirement. In an acquisition/divestiture process, the focus is on the presence (or absence) of *major* regulatory program components. Minor technical violations are generally not of interest, since resolution is of minor economic impact and usually will not affect the viability of the facility. Because of the nature of the findings, which could support future penalty or other enforcement actions, it is important to discuss procedures first. Preliminary oral reports to allow corrective measures to be implemented immediately are often desirable.

An action plan can be developed that sets forth prioritized tasks designed to address identified program deficiencies. If a high level of noncompliance is noted, overall EHS program management may be deficient, and a management appraisal (see below) may be appropriate.

## Who Can Do It?

Senior regulatory specialists are required to perform compliance assessments. Acquisition/divestiture projects are usually fast-track assignments; therefore, broad-based knowledge and extensive experience are required to quickly translate large volumes of data into succinct summaries. Junior personnel will generally lack the experience and judgment necessary to perform in this manner.

## MANAGEMENT ASSESSMENT

### Why You Do It?

Management assessments are special-purpose audits which are designed to assess the EHS management structure of the organization, and assure that the system can operate and respond effectively to meet the integrated requirements of EHS management. When assessing the management structure of an organization, the focus is on people and systems. The main questions to be answered are: (1) how are people communicating and are they working together in an efficient manner toward common identified goals?; and (2) do the current management systems help or hinder the workforce? Companies perform these studies to improve overall corporate efficiency in anticipating and complying with regulatory requirements, and in reducing overall EHS liabilities. In an acquisition process, a management assessment will additionally focus on the following:

- Who holds the key EHS positions within the company to be acquired, what are their skills and levels of competence, and do those skills and competencies mesh with personnel within the purchasing organization?

- What are the long-term EHS management goals and strategies, and do they coincide with those of the purchasing organization?
- In general, do the organizations share the same values and EHS management styles?

## Collecting Information

A thorough evaluation of management systems is usually not possible, given the time frame of an acquisition transaction. However, if a management assessment is desired, it may be performed concurrently with a compliance assessment. Many compliance issues will have a root cause based in one or more management systems. Delving deeply into compliance issues will frequently lead to at least a limited evaluation of a discrete management system. The following systems can be evaluated, relying heavily on extensive interviews with key management and line personnel.

- Internal standards, policies, and procedures;
- Roles and responsibilities;
- Communication;
- Organization, staffing, and structure;
- Business planning and strategy;
- Project and program planning;
- Performance standards and indicators;
- Regulatory tracking;
- Liability and risk management;
- Compliance management;
- Management reporting systems.

Resulting findings will identify management dysfunctions and recommend possible procedural, structural, or organizational changes to correct problems. An alternative to the above list would be to use an environmental management standard, such as those published and/or proposed by the International Standards Organization (ISO).

## Who Can Do It?

Senior management consulting experts are required to perform a management assessment. Since, in an acquisition transaction, a management assessment is usually performed concurrently with a compliance assessment, many efficiencies can be gained if the same people perform both tasks. Evaluator duality can be achieved through the use of compliance experts who also have extensive and diverse management experience.

## ADDITIONAL TYPES OF INVETIGATIONS AND INVESTIGATION STRATEGIES

In addition to mixing and matching components from the investigations described above, the following special strategies should also be considered in an acquisition process:

1.  Select cross sections of representative sites to be inspected. All properties in a multi-site acquisition need not be inspected if a reasonable representative sample of sites can be selected.
2.  Assess off-site waste disposal liability. If the structure of the transaction results in the purchaser acquiring or otherwise assuming liability for the off-site waste disposal practices or former facilities of the acquired company, then an assessment of those practices and

facilities is in order. This assessment would be a combination contamination, compliance and management assessment, focusing on historical waste handling and disposal practices and practices at the closed facilities.

3.  Wetlands and environmentally sensitive features. If real property is being acquired for development or an expansion project is planned for a key operating facility, an investigation of the presence of wetlands or other environmentally sensitive features that might affect developability is important. In real estate acquisitions, such investigations are typically part of a Phase I and are often limited to the type of investigation pursued in a Phase I contamination investigation. For example, for a Phase I study regarding wetlands, one would consult records, such as soil surveys and the Army Corps' National Wetlands inventory, and conduct a field walk to allow rough identification of the extent of the wetlands. A more thorough delineation may later be necessary. A Phase I may also look to public records for reports of federal or state threatened or endangered species. A certification program is being developed for wetlands delineations, this work should be done by biologists/ecologists/environmental specialists or others with training in plant and animal identification, soils, and hydrology.

4.  Include special nonregulatory issues in a compliance assessment. The following issues, although not totally driven by regulations, can be effectively evaluated under the auspices of a compliance assessment.

    *   Pollution prevention/waste minimization;
    *   Product life cycle and cost accounting;

- Process safety and emergency management (including toll contractors).

5. Develop a strategy to modify policies, procedures, and operating practices immediately upon assumption of ownership and operating responsibilities. Modifying policies, procedures, and practices in this manner clearly establishes a line of demarcation between old management and new, and will make it easier to segregate responsibilities for issues which may overlap regimes.

## CONCLUSION

In closing, two key encompassing concepts to be applied to "environmental investigations" should be stressed:

(1) There is no standard approach for environmental investigations in transactions.

The investigation will vary from transaction to transaction, depending on the type of transaction, the parties' goals, type of property/facility, time constraints, and budget constraints. Minimally sophisticated transactions involving multiple sites or industrial operations will usually require a customized combination of the three basic types of investigations discussed above.

(2) The roles and responsibilities of investigation participants are multiple and overlapping.

The entire project team must agree upon and understand their respective roles and responsibilities. While certain professionals, due to training and background, may be more appropriate in certain roles, individual experience and capabilities vary considerably. Once roles and responsibilities are agreed upon, team members must frequently and thoroughly communicate with each other, explaining and discussing the multidisciplinary significance of the information and issues for which they are responsible. The best approach to assure a successful investigation is to have talented people, in appropriate roles, effectively communicating.

## ENDNOTES

[1]  The standard for environmental investigation developed by the American Society for Testing and Materials ("ASTM") is an example of procedures that include such flexibility.

[2]  Environmental consent decrees and orders also typically require some notification regarding property transfers and changes of control.

[3]  It is equally possible that bringing a company into compliance could have benefits. Waste reduction often reduce operational costs.

[4]  ISO 14000 standards are being developed by the International Organization for standardization. Standards are being developed for environmental management systems, environmental audits, terminology, performance evaluations, labeling, and life cycle analysis.

5    This is not meant to suggest that the defense is irrelevant for all purposes. It would be important to an owner seeking contribution from the person causing the contamination and would also be important to an owner holding the property for productive use (assuming remediation would not interfere with the productive use).

6    This is not always the case. Tank laws, for example, impose obligations on consultants.

7    This is not meant to suggest that the defense is irrelevant for all purposes. It would be important to a party seeking contribution from the person causing the contamination and would also be important to an owner holding the property for productive use (assuming remediation would not interfere with the productive use).

# 14

## ROLE OF AUDITING

*Paul Pizzi and Mike Henke, Pilko & Associates, Inc.*

### INTRODUCTION

Audits are a classical management tool that involve the application of an investigative, analytical, comparative process to variables considered critical to a company's success. They were originally designed to exert control over company operations and correct negative deviations from minimum standards. The most familiar of audits relate to financial matters. However, audits have been used extensively during the past 15 to 20 years to evaluate environmental, health and safety (EH&S) issues manufacturing industries and their scopes have expanded to include identifying operations improvements, in addition to correcting negative deviations.

The use of auditing is an important element of any EH&S management program and it has been endorsed by organizations such as the United States Environmental Protection Agency, the International Chamber of Commerce, the United States Department of Justice, The Environmental Auditing Roundtable, and the Global Environmental Management Initiative. Environmental management standards being developed (*e.g.*, ISO environmental auditing standards) or already in place also describe key roles for audits in comprehensive EHS management programs.

Audits can have many benefits. A key assumption behind the use of audits is that EH&S performance will be improved as a result of the application of the audit process. The mere process of satisfying the information needs of an audit can improve the facility's or company's ability to demonstrate compliance with regulatory requirements. Audits can also identify and evaluate risks and foster effective management of those issues. Awareness of EHS issues is generally enhanced as a result of audits. Information generated can to satisfy certain public reporting requirements or to demonstrate EHS commit various constituencies. Finally, in a defensive vein, the conscientious application of programs can be used as a negotiating tool to reduce penalties in the event of a regulatory violation and in cases of environmental or safety incidents.

## Historical Perspective

Environmental auditing in the U.S. started as early as the late 1970s, with the impetus provided by both the Securities and Exchange Commission (SEC) and the Environmental Protection Agency (EPA). The SEC, under consent orders, required U.S. Steel (1977) and Allied Chemical (1979) to undertake a comprehensive companywide audit program to determine their environmental liabilities, which the SEC felt were being under-reported in the company financial statements.In addition, the rapid rise of environmental regulations promulgated by EPA in the Mid-70s through the early-80s, with their costly penalties for noncompliance, motivated companies to establish audit programs to assure compliance. Programs continued to develop in the 1980s, with some larger companies creating an auditing group at the corporate level, staffed by a full-time audit manager and dedicated staff. The programs became more formalized and consistent, and today nearly every major corporation has a well-defined EHS auditing system.

In 1986, EPA published an environmental auditing policy statement, which stated that they would continue to propose environmental auditing provisions in certain consent decrees, affirmed the value of environmental self-assessments, and established a set of guidelines for effective auditing programs. In April of 1995, EPA published an interim policy designed to provide incentives for the regulated community to conduct voluntary compliance audits and to disclose and correct any violations discovered during those audits.

## DRIVING FORCES FOR CHANGE

Today, major transitions are occurring in EH&S auditing programs, spurred by several driving forces:

- Federal EH&S legislation and regulations continue to proliferate and grow in complexity;
- States and municipalities are becoming more sophisticated with regard to EH&S issues, resulting in more and tougher state and local regulations;
- The EPA continues to increase the aggressiveness of its enforcement initiative. Many companies have already felt the sting of audits from the recently created National Enforcement Investigation Center (NEIC) in Denver, Colorado;
- Senior management has serious concerns about certain provisions of the Clean Air Act Amendments and the potential for criminal liability resulting from EPA's Sentencing Guidelines. The Department of Justice's Responsible Corporate Officer Doctrine and

beefed-up staff, and the environmental focus of the Clinton administration mandate a very close watch over EH&S compliance.

All of these factors are driving industry toward a renewed interest in *compliance verification* at the "nitty gritty" level. Yet constant regulatory changes make compliance something of a moving target, which leads to the second major focus today: a heavy emphasis on *management systems*. Companies want to ensure ongoing compliance by having strong management systems and good operating practices in place and functioning throughout the organization.

## CHALLENGE OF MAINTAINING COMPLIANCE

Part of the problem in ensuring compliance is finding the personnel to do it. A significant number of companies in a wide spectrum of industries are reorganizing, reengineering, or "right-sizing." While compliance issues increase in both number and complexity, qualified resources are becoming increasingly limited and strained.

The second problem is finding a workable structure for the audit management system: How can senior management ensure that every facility is in compliance?

To address both of these concerns, many companies have supplemented their audit staff with third parties. Outsourcing in audit programs can be grouped into four general categories:

- *Additional " hands."* Using a third party as a team member or a team leader can help meet overall resource requirements.

- *Specific expertise.* With limited resources, it has become more difficult to find internal staff with knowledge of specific regulations who have also been trained in auditing. Outside resources can often fill this gap.

- *Program development and management.* Some third parties have auditing experience in many different venues, and therefore can often offer invaluable assistance in establishing a corporatewide audit structure.

- *Quality assurance.* These top-level reviews are becoming commonplace. Typically, companies hire a third party on a periodic basis (two- to four- year intervals are most common) to review the entire audit process. The third party can assure senior management that the audit program is functioning effectively, that the objectives of the audit program are being met, and that the objectives remain valid given the changes in regulatory requirements, stakeholder demands, company vision, etc.

## EMERGING TRENDS

Two important trends are emerging which will shape auditing throughout the 1990s. The first is a *layered* approach to auditing, and the second is an increasing interest in *auditing metrics*, or cost/benefit analyses of audits and auditing programs. Each of these trends is discussed below.

### Layered Approach to Auditing

In order to manage detailed compliance issues better and provide the necessary assurance to senior management, many companies with single-focus audit programs are restructuring to a layered system. While the specific program varies according to the organization and its degree of centralization, there are three basic levels in a layered audit program:

- *self audits* at the facility level;
- *compliance assurance* audits at the corporate or divisional level;
- *oversight* audits at the corporate level.

## Self Audits

These audit programs are being driven by the EPA's enforcement focus. In self audits, each operating facility audits itself for regulatory compliance, usually on a continuing basis throughout the year. This is typically a very detailed and rigorous audit that encompasses state and local, as well as federal requirements. Self audits are conducted on a real-time basis, with the ultimate objective of flagging noncompliance areas through daily monitoring.

There is considerable flexibility in structuring a self-audit program. Audits can be organized according to specific processes within the facility. This type of organization lends itself to refineries, chemical plants, and other complex manufacturing operations.

Self-audits may also be structured more broadly, where various parts of the audit cover air, water, waste systems, and so on, across the entire facility.

By definition, self-audits focus almost entirely on regulatory and operational details typically contained in voluminous self-audit manuals that include policies, protocols, and checklists. Keeping these huge manuals up-to-date in the face of continually changing regulations is a daunting task, and some companies are launching efforts to streamline and computerize the process.

In addition to ensuring regulatory compliance, self audits serve a training function in many facilities: By serving on audit teams, plant manufacturing personnel gain an increased awareness of important EH&S issues in their operations.

## *Compliance Assurance Audits*

The second audit level is the compliance assurance program, which is an independent assessment of the compliance status of a company's facilities. This type of program developed in the 1980s and is now considered a fairly traditional approach. It grew out of a push to assess regulatory compliance at each facility and to provide compliance status information to senior management.

Most compliance assurance programs are conducted at the division headquarters or the corporate level. The compliance assurance audit includes both operating facilities as well as nonoperating locations such as research & development laboratories, storage terminals, or office buildings. Compliance assurance programs include checklists for regulatory compliance, and they often review the management systems in place within each facility.

## *Oversight Audits*

While compliance assurance audits at the corporate or divisional level do much to ensure that facility self audits remain on track, they may not provide the complete assurance sought by senior management who retain the burden of criminal liability in cases of negligence. For this reason, a third audit level is now gaining acceptance.

The third level is an "oversight" program conducted by corporate or division headquarters, and it focuses on management systems and audit processes rather than detailed compliance issues. Specifically, the oversight audit looks at compliance assurance programs (or self audits if there is no compliance assurance program in place) to assure senior management that these programs are operating effectively and meeting established corporate guidelines and/or standards.

Tenneco offers one example of such a program. Michele Malloy, Tenneco's Director of Environmental Affairs, says that Tenneco's auditing program has evolved several times since it was started in 1978: "Our audit program began as a compliance assurance program conducted by corporate staff. Later, the divisions took responsibility for compliance assurance, and the corporate program became an oversight audit of the divisional programs.

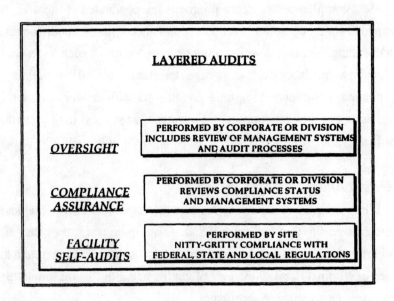

**LAYERED AUDITS**

*OVERSIGHT*
PERFORMED BY CORPORATE OR DIVISION
INCLUDES REVIEW OF MANAGEMENT SYSTEMS
AND AUDIT PROCESSES

*COMPLIANCE ASSURANCE*
PERFORMED BY CORPORATE OR DIVISION
REVIEWS COMPLIANCE STATUS
AND MANAGEMENT SYSTEMS

*FACILITY SELF-AUDITS*
PERFORMED BY SITE
NITTY-GRITTY COMPLIANCE WITH
FEDERAL, STATE AND LOCAL REGULATIONS

Recently, we have expanded the function of the corporate oversight audit to include proactive risk assessment of environmental issues at targeted facilities that may or may not fall under current regulatory guidelines. This layered approach works well and appears to provide the right balance between oversight and facility reviews."

## Auditing Metrics

The idea of "auditing metrics" has been generating much interest recently. In the current economic climate, every aspect of a business must justify its impact on the bottom line. Auditing is a function performed solely to assess compliance, reduce risk, and prevent loss—how can its efficiency be evaluated?

Many environmental auditing program managers are being asked to come up with some measurement of efficiency that can be reduced to numbers and monitored. Complicating the problem further are two facets of auditing cost/benefit analysis

- Analysis of the audit: Did it check the right things and reduce the right risks so that management is not paying "too much" to prevent accidents that have an extremely small probability of occurring?
- Analysis of the auditing program: Are audits being carried out often enough but not too often? Are the mechanics of conducting the program as cost-efficient as possible?

The metrics concept will be most easily applied to auditing programs with a large existing database, rather than to a new program from the ground up.

## TYPES OF AUDIT PROCESSES

EHS audits normally devote some attention to identifying conformance (i.e., compliance) with regulatory and policy requirements. More comprehensive programs also address the design and implementation of management systems to ensure responsible operations. Other audits strive to identify activities that represent an unacceptable risk of potential losses for the company.

### Compliance Audit

The most common type of EHS audit relates to the operational conformance with standards established by entities such as federal, state, and local governments (*i.e.*, laws and regulations); company management (i.e., policies and procedures); and trade associations (*e.g.*, Responsible Care, STEP, etc.) These audits tend to be rather black-or-white affairs that address specific requirements and avoid less clear-cut issues such as good practices or risk assessment. Although it is important to check compliance status, these types of audits represent, at best, a "snapshot in time" because they tend to focus on the past or present. As financial advisors are fond of saying, "past performance is not a guarantee of future returns."

### Management Systems Audit

A second type of EHS audit acknowledges that the primary responsibility for compliance should rest with the line organization, but that management has a responsibility to ensure that adequate company resources are being properly applied to EHS issues. Under this type of audit, the primary focus is to evaluate the scope and effectiveness of systems employed by the facilities to manage EHS affairs. Compliance status is reviewed as an indication of the performance of these management systems, but compliance determination is not the sole, or even the main, objective.

The management systems audits may not be adequate to meet the needs of facility management if they have a genuine need for guidance or interpretation of the requirements. Under this scenario, plant managers often complain that the audit program has missed a detail that ultimately resulted in a citation or fine. Another potential problem with these types of audits is that clear-cut standards have not been developed for EHS management systems. As a result, questions of management style may be highly debated and not all corporate cultures are receptive to this approach.

## Risk Reduction Audit

A third type of audit, and one that is often very sensitive and misunderstood, involves attempts to evaluate the risk of losses to a company. Losses can be defined in a number of ways such as fines, civil penalties, lost-time incidents, fatalities, negative impacts to company image, or loss of market share. Generally speaking, without a significant coordinating effort, no two people will define losses in the same way. A second potential source of confusion can relate to estimating the probability that an event will occur or estimating the size of a potential loss. Overall, there may be legal concerns about employing any process that contemplates potential losses and attempts to estimate them. Despite these difficulties, companies that have a good deal of confidence in their ability to manage will forge ahead and work through the difficulties in the belief that the process will reduce overall risks.

## Design of an Audit Program

The exact design of an auditing program should be the product of company culture, business objectives and the specific objectives of the EHS management program. With some companies it makes sense to use multiple layers of audits involving self-audits as the primary check for compliance and overview audits to evaluate management systems and risks. Other companies use overlapping audit programs (i.e., business unit and corporate audits) to conduct audits that may have redundant objectives. Still other companies set audit system standards at the corporate level and rely upon business units to develop and implement conforming programs.

Regardless of the specific program design selected, it is critical for the company to clearly define the key customers for the audit program and determine their needs. Auditors must be selected with care to ensure their independence (i.e., absence of bias) and adequately trained to ensure competence. Protocols and procedures for the audits should be defined in writing to provide reproducibility and a clear understanding of the meaning and significance of the findings. Findings should be documented and procedures should be in place to facilitate the implementation of corrective actions or improvements. Finally, the audit process should be periodically reviewed to ensure that it is properly designed and implemented to meet dynamic business objectives and changing regulatory requirements.

## AUDIT RISK

It is important to recognize that audits do not provide guarantees of compliance, risk reduction, or effective EHS management. Financial auditors refer to this concept as "audit risk." Basically, audit risk is the potential for an audit to fail to identify an item that it is designed to find.

Audit risk arises out of several factors. First, economic considerations make it infeasible for a company to devote the time and resources to examine every document, interview every individual, or inspect every piece of equipment. Second, audit processes are subject to the law of diminishing returns. Consequently, effective audits rely upon statistical inferences to draw general conclusions based upon analyses of samples.

Another factor affecting audit risk is the level of regulation and the complexity of the operation. This factor, which is known as inherent risk, says that the likelihood of failing to identify a nonconformance is directly proportional to the number of regulations and the complexity of regulations applicable to a particular operation. For instance, for a given scope and level of resource commitment, it is more likely that a noncompliance would go undetected by an audit at a refinery than at a marketing terminal. To avoid varying degrees of audit risk, companies often increase the size of an audit team used at a complex facility, devote more time to the audit, or reduce the scope.

A third factor impacting audit risk is the extent and effectiveness of systems employed by the operating facility to manage EHS issues on a daily basis. Generally speaking, the more sophisticated the EHS management system used, the more reliable is the information generated by the system. As a result, the audit team may be able to rely upon a summary document prepared by the facility rather than conducting an independent review of all the documents used to prepare the summary. However, the team should evaluate the procedures used to prepare the summary.

Because audit risk exists, companies must develop and implement an audit process that provides reasonable assurance that the objectives of the audit will be met. This level of assurance is a subjective concept and cannot be precisely measured.

## COMMON CHARACTERISTICS OF AN EH&S PROGRAM

Although audit programs can vary from one company to the next, effective EHS audit programs have several common characteristics that enable them to function successfully. These characteristics include

- Compatible approaches and objectives;
- Top management support;
- Auditor proficiency;
- Independence;
- Due professional care;
- Explicit program description;
- Systematic plans and procedures;
- Planned and supervised field work;
- Quality assurance;
- Records management;
- Clear and appropriate;
- Commitment to follow-up.

### Compatible Approach and Objectives

When a company establishes an environmental management function, an important cornerstone of the function is the development of an approach for compliance assurance, including audits. These approaches generally fall into three categories: self-audits, centralized audits, and decentralized audits with centralized oversight. The selection of the proper approach depends on the resources available for conducting the audits, the capabilities of the audit group, and the culture of the organization. Each of these approaches have distinct characteristics and it is important that the approach matches the objectives of the audit program and the culture of the organization.

Self-audits are typically driven by the establishment of checklists and protocols which are distributed to all of the operation functions and facilities. Environmental personnel at the operating facilities are then required to complete the checklist by a particular deadline and establish a corrective action plan. This approach is often helpful for enhancing the awareness of environmental issues throughout the organization, but it relies solely on the diligence and judgement of the individual auditors. This approach offers the greatest potential for inconsistent interpretation of environmental standards and application of the checklists. The checklists must be updated frequently to reflect changes in requirements.

Centralized audit programs are typically administered by a full-time audit manager or by an environmental manager on a part-time basis. Teams of experienced individuals are established on a rotating basis to conduct the audits. Personnel for the teams are gathered from different parts of the organization for a period of usually less than one week. Because the team members come from different backgrounds and have different areas of expertise it is important to train the team members to assure good understanding of the audit's objectives and scope, auditing techniques, and the operations being reviewed.

Decentralized audit programs allow for adjustments that may be necessary to accurately review operations in different business units or in different geographic jurisdictions. Each business unit typically establishes an audit program, but to ensure communication to the highest levels of the organization and to enable directors and officers to manage environmental risks, a centralized program of oversight is established. The central audit group may establish minimum requirements for the individual audit programs and conduct internal quality assurance reviews on a regular basis.

## Top Management Support

The U.S. EPA indicates that top management support for environmental auditing is critical to the continuing success of an audit program. The U.S Department of Justice also specifies that corporate compliance programs must designate a specific, high-level individual to have primary responsibility for ensuring implementation of a compliance program and for compliance with the corporation's policies and procedures. A review of the company's organization chart is often a good indicator of the degree of commitment to environmental management.

Environmental audit programs must reside at the proper location in the organization to foster real improvement in environmental management. In general, this means that the person responsible for the program should report to at least the vice presidential level. In the absence of a prominent position in the company, strong reporting mechanisms should exist to ensure that major issues are properly considered and addressed in a timely fashion.

Management support may be demonstrated by a written policy articulating upper management support for the auditing program and for compliance with pertinent requirements. It may also be demonstrated through routine inclusion of environmental topics in management meetings or staff meetings, and through integration of environmental performance in the employee performance review process.

It is also useful to investigate other criteria to evaluate management support. How have conflicts regarding environmental management and other priorities been resolved? Have operations ever been shutdown or curtailed when compliance could not be achieved? Have appropriate resources been allocated to environmental management projects? Does management frequently emphasize the importance of environmental performance in newsletters or other communications? Are awareness training programs in place? Are corrective actions implemented in timely fashion?

## Auditor Proficiency

Quality assurance reviews seek to determine whether appropriate personnel are selected to conduct environmental audits.

The specific staffing of an audit team depends on the objectives, scope, and methodology of the audit program as well as the culture of the organization. The staffing must be adequate to conduct the audit within a reasonable time frame that minimizes disruption of facility operations, while providing reasonable assurance that the objectives of the audits will be met.

In order to consistently and reliably execute audits, the members of the audit teams must possess adequate technical knowledge, experience, and skills. Auditors should have a good working knowledge of auditing process procedures and techniques, regulatory requirement, characteristics of management systems, environmental protection systems, facility operations, and potential environmental hazards. To effectively execute an audit, the auditors must have good interpersonal and communication skills, be able to schedule and plan work, and possess good analytical skills to evaluate information collected. Finally, the auditors should also understand the overall objectives of the audit program and the specific objectives of the audit. Most companies ensure the proficiency of audit team members through the establishment of training programs and minimum qualifications for audit team members and team leaders.

Many companies borrow personnel from other parts of the organization to complete the desired number of audits and ensure that the teams have the right blend of experience and expertise. If the audit staff or audit team members are inexperienced or lack the necessary skills, training is needed to ensure the effectiveness of the program. A written curriculum and formal training are best. Otherwise, inconsistencies may develop because of differing priorities or interpretations, as audits are conducted on an ongoing basis.

Perhaps the most important characteristic of effective team leaders is an ability to credibly work with a wide range of individuals from the board room to the control room. Many of the issues identified during audits require technical solutions, but their significance must be interpreted and credibly communicated in a business, financial, or strategic context. Selecting auditors with experience in the line organization is one of the best ways to develop this credibility.

## Independence

To ensure the objectivity of the audit process, it is important for auditors to be sufficiently detached from the activity being studied. Bias may arise out of organizational considerations (*e.g.*, reluctance to criticize a plant manager for whom you may eventually work) or personal experience (*e.g.*, being asked to evaluate a plan that the auditor previously developed). The audit program should be designed to avoid conflicts of interest and eliminate pressure on auditors to influence their findings. The most sophisticated audit programs will include mechanisms to identify potential conflicts and communicate potential conflicts to the client.

In practice, it is very difficult to assemble a completely independent team of auditors to conduct internal audits. To facilitate the swift and effective completion of an audit, a company requires a minimum level of operational or facility knowledge for the audit teams. This is accomplished by staffing the team with personnel from similar facilities (*e.g.*, refinery staff are selected to audit a refinery) and this practice introduces a potential for influence over the audit findings. The audit program must properly balance this knowledge and independence of the audit team. Ultimately, the client of the audit program must make the final decision with regard to the degree of independence required to accomplish the program objectives.

## Due Professional Care

The requirement for due care imposes a responsibility upon each member of an audit team to conscientiously apply diligence and skill in performing an audit and to consistently apply audit standards and procedures. The intent is to ensure the accuracy, consistency, and objectivity of the audit. As such, auditors must apply their diligence and skill in the same manner that other competent, reasonably prudent and knowledgeable auditors would apply them under the same or similar circumstances. To assist the auditors, the auditing program should include provisions whereby authoritative interpretations can be obtained during an audit when standards conflict or appear vague.

Opinions developed during the audits should be supported by reasonable and sufficient evidence. Objective findings should be based upon observed, measurable, and verifiable evidence.

## Explicit Program Description

The U.S. EPA recommends that companies should explicitly define the objectives and scope of their audit programs, along with the resources to be employed and the frequency of the audit.

As discussed previously, it is critically important that the objectives of an audit program are consistent with a company's culture and the company's strategic business objectives. However, for audit programs to be most effective, their objectives must also be clearly understood and communicated throughout the organization.

Typical objectives of audit programs include assurance of compliance with regulations, policies, and standards; confirmation that management programs are functioning effectively; evaluation of hazards; and reduction of risk, or liabilities. Other secondary objectives may include increased environmental awareness, collection of data, training of personnel, and reduction in cost.

If the objectives of the program are not clearly defined or are ineffectively communicated, audits are often conducted inconsistently and incompletely. It may also be difficult to obtain the necessary internal resources on loan from other areas of the company and, ultimately, little real improvement in environmental performance is sustained.

The written audit should describe the scope of the audit program, including which facilities are to be audited, what topics will be covered, and what standards will be used to measure or evaluate performance. In the 1990s, it may no longer be appropriate to limit the scope of an audit program to only company-owned domestic manufacturing facilities. Significant hazards or liabilities may arise out of the operation of foreign plants, joint venture facilities, distribution terminals, toll processors, and other activities. Similarly, it is often helpful to evaluate commercial waste management facilities prior to first use and periodically thereafter.

With the globalization of many audit programs, it is also important to spell out the standards or guidelines that will be used by the audit teams as they develop their findings. Confusion often arises because regulatory requirements, and customs differ throughout the world. Guidance regarding the standards to be used will help avoid confusion and will focus the environmental management efforts of operating personnel. The audit process should ensure that information about pertinent corporate policies, permits, and regulations is readily available to the auditors. Most often checklists or protocols are used to provide this guidance. However additional resources (*i.e.*, newsletters, regulatory summaries, databases, procedures manuals, etc.) may also be utilized.

Finally, the definition of scope should establish which environmental topics (*i.e.*, air emissions, waste management, wastewater treatment, etc.) will be addressed and the time period to be covered by the audits to ensure that auditors focus their investigation efforts on the period that will be most useful to the company.

Audit programs can also run more smoothly if the mechanisms under which resources, mainly personnel and money, will be made available for audit teams are spelled out in the overall program description.

## Systematic Plans and Procedures

As a means of ensuring consistency, quality, and thoroughness, the audits must be adequately planned and the field activities of the audit team must be properly supervised. Audit planning involves developing an overall strategy for the scope and conduct of the audit. A clear understanding of the audit strategy is critical because the nature, extent, and timing of audits may vary with the size and complexity of the facilities involved, and the audit team's experience with the facility and knowledge of the facility's operation.

A written audit program description, including protocols and procedures, helps ensure good understanding of the work to be done. This program description also helps ensure thorough and consistent preparation for audits, effective execution of the field investigation and appropriate documentation of findings.

Systematic audit planning processes and procedures include the use of protocols, checklists and guidelines that are consistent with the audit scope and objectives. These documents provide a clear outline of the methodology used to conduct the audits, manage working papers, and document the findings. Examples of these procedures and guidelines include:

- Selection and scheduling of sites for audits;
- Production, maintenance, and use of protocols, checklists, and questionnaires;
- Selection of audit team members and leaders;
- Pre-audit information collection;
- Preparation and management of working papers;
- Preparation, review, and distribution of reports;
- Development of corrective action plans;
- Records management;
- Quality assurance;

Typical compliance areas to evaluate in an environmental audit are as follows:

- Wastewater discharges;
- PCB's;
- Air emissions;
- Drinking water;
- Hazardous waste;
- Use of pesticides;
- Oil spill control;
- Underground storage tanks;
- Solid waste;
- TSCA requirements.

## Planned and Supervised Field Work

Audit field work must be adequately planned and implemented. Since conducting an audit typically involves a team effort, the activities of team members must be properly supervised to promote efficiency and consistency with the audit objectives.

Team members must have access to, and be familiar with, the audit protocols and any plans uniquely developed for the specific audit. The audit process should include guidelines for auditors to help them collect and evaluate evidential matters. Testing and sampling techniques should be agreed upon in advance, to the extent practical, but flexibility is needed to allow expansion or alteration of these techniques, if warranted.

Supervision and leadership for the audit teams is typically provided by a team leader who possesses particular experience and coordination skills. The team leader should ensure that the auditors gather information that will support the objectives of the audits. In order to be most compelling, the information used to support audit findings should be valid, relevant, accurate, and sufficient to afford a reasonable basis for opinions. The amounts and kinds of evidence required to support an opinion are matters for the audit team to determine by exercising professional judgment after careful analysis of a case's particular circumstances. However, the evidence should be convincing enough that a prudent, informed person would be likely to reach the same conclusions.

## Quality Assurance

The U.S. EPA recommends that quality assurance procedures be included in audit programs to assure the accuracy and thoroughness of environmental audits. The assurance may be accomplished through supervision of audit teams, independent internal reviews, external reviews, or a combination of these approaches. Overall audit quality is determined by the extent to which they are conducted in conformance with auditing standards and the objectives and scope outlined in the program description. Checks of individual audit quality should be integrated into auditing procedures to ensure that findings are consistent with evidence recorded by the auditors and that the findings are reliably communicated in written reports.

## Records Management

Audit teams generate working papers that describe the conduct of audits, the operation of facilities, and the status of environmental issues. Similarly, audit reports contain observations, conclusions, and recommendations regarding environmental issues. These documents may contain sensitive information and must be treated with care.

The audit program should include a written procedure outlining the preparation, review, distribution, and storage of working papers and reports generated during an audit. The procedure describes the purpose, content, and use of the working papers along with the interrelationships among the working papers, questionnaires, protocols, and reports. Consistency is fostered when guidance is provided to the auditors on the format of the working papers and procedures for obtaining any necessary reviews or approval. The procedure should also address access to records, retention periods, and the storage location for the documents associated with the audit program.

## Clear and Appropriate Reporting

The findings of environmental audits must be properly documented. To ensure this, audit processes should include specific procedures for promptly and candidly documenting the findings of individual audits and communicating them to appropriate levels of management. Reports should be based primarily upon factual, objective observations and should avoid speculation, subjective opinions, and sensitive language.

Two types of reports are typically used: descriptive reports and exception reports. Descriptive reports include descriptions of the operations, along with discussions of compliance status, hazards, risks, and potential liabilities. Exception reports indicate only those situations where deviations (positive and negative) exist from requirements or good practice.

In general, written audit reports should include descriptions of the audit scope and conduct as well as discussions of the audit results and conclusions. Findings should be supported with relevant and accurate information that is sufficient to satisfy the objectives of the audit. Depending upon the culture of the organization and the objectives of the audit, recommendations may be included in the reports and an auditor's opinion of the overall status of the facility may be included.

Procedures for preparing audit reports should outline the role of various team members in preparing the written report and assign primary responsibility. Guidelines should specify deadlines for the preparation of a draft report and procedures for review, finalization, and distribution.

Written reports should be distributed to key personnel who can commit the resources of the company to improve compliance status, reduce risks, or achieve other objectives. Adequate opportunities should be provided for review of a draft report to ensure accurate description of operations and proper characterization of risks. If the report contains recommendations, they should be flexible enough to allow facility personnel some latitude in determining how improvements will be implemented, but they should be definitive enough to ensure continuous improvement.

Regardless of the type of written report employed, the findings and recommendations should be prioritized so that resources can be properly allocated to address the most critical issues on a timely basis. It is often helpful to organize the findings and recommendations into several categories depending on their importance and the level of authority needed to address them. These categories should be highlighted in the written report to enhance understanding and for the convenience of the reader.

## Commitment to Follow-Up

Audit programs are an important component of a company's overall environmental management program and can help promote real improvements in environmental performance; but improvements cannot be implemented on a timely and consistent basis without a concrete, systematic plan for following up on the audit findings and implementing corrective actions.

The follow-up may be the most important aspect of an audit program. Simply stated, it is not worth conducting audits if the company is not committed to following up on findings. Liabilities can multiply and adverse publicity can result through inaction. Companies may also increase the likelihood that they will face criminal prosecution if they lack a plan for dealing with violations detected during internal environmental audits.

To ensure progress, it is essential to clearly delegate responsibility for corrective action to an appropriate person or group. In most companies the line organization is responsible for implementing improvements. To encourage positive action, action plans are normally prepared on a timely basis to outline the action to be taken, assign responsibility, and establish a tentative deadline for completion. Action plans are then monitored on a regular basis.

Organizational sentencing guidelines adopted by the U.S. Department of Justice also provide compelling reasons for companies to follow up on audit findings. Criteria used to determine if a company has implemented a compliance program (and, therefore, qualifies for favorable sentencing adjustments) include consistent enforcement of compliance policies through disciplinary measures and appropriate response to misconduct to prevent recurrence.

Management's commitment to the follow-up should be demonstrated explicitly in writing. This commitment often manifests itself in the form of procedures outlining the development, selection, implementation and monitoring of corrective actions. The objective of these procedures is to demonstrate the timely completion of improvements designed to assure compliance and manage risks.

## ATTORNEY-CLIENT PRIVILEGE

As companies conduct audits to help them attain full compliance, they most likely will uncover instances of non-compliance that could be used by citizens or regulatory personnel to bring litigation against the company. They are then faced with a decision—should they try to protect this information from discovery efforts, that will involve a strict procedure to contain the information, or should they share this information within the company to assist in meeting compliance at their other facilities?

The answer is not self-evident. However, if a company chooses to prevent non-compliance from discovery, there are certain criteria that must be adhered to:

- The audit must be set up and conducted under the direction of the company's attorney;

- The audit information is used by the attorney to provide legal advice on compliance status and potential liabilities;
- The company must handle the information with strict confidentiality.

Although four states (Colorado, Indiana, Kentucky, and Oregon), have passed laws that extend a limited privelege to information contained in voluntary audits, the U.S. EPA has consistently opposed the state legislative enactments creating an environmental audit privelege. EPA and the Department of Justice, however, will take the fact that a company has a self-auditing program in place as a mitigating factor in relation to penalties under the *Draft Corporate Sentencing Guidelines for Environmental Violations*.

For those companies who are just beginning a formal self-auditing program, attornehy-client privileged audits may be useful until the company's degree of compliance has been evaluated and steps taken to correct deficiencies. As a program matures and the company become more proactive in it's environmental efforts, the attorney-client privelege may be a barrier to utilizing information from audits for companywide improvement continuous improvement.

Before undertaking an environmental self-audit program, the company's attorney should be consulted to obtain advice on maximizing confidentiality of audit results while allowing flexibility to utilize learnings for self-improvement.

## FUTURE FORECASTS

The auditing programs of U.S. companies will become still more formalized and more consistent with the advent of International Environmental Management Standards from the ISO 14000 process over the next few years. The European Community has already adopted a "voluntary" Eco-audit scheme that includes some elements of auditor certification and public reporting.

The growing complexity of regulations, coupled with the demands for more information about a company's environmental performance will require the environmental auditing process to continue developing and maturing, create the need for certification of auditors to provide quality reviews, and ultimately open the door to public reporting of environmental audit results.

# 15

---

# BENCHMARKING

*Ronald W. Michaud, Pilko & Associates, Inc.*

## HISTORICAL PERSPECTIVE

Perhaps the greening of America's corporations had its roots in the early 1970s, after the first Earth Day in April 1970. Or perhaps, it was in response to the formation of the Environmental Protection Agency. Or, maybe it was in reaction to the legislative fervor that led to passage of a significant number of environmental statutes during the 70s and 80s, and the subsequent regulation that infringed on industry's flexibility to conduct business as usual. Perhaps it was a singular event, such as Bhopal, which heightened industry's awareness that a disaster could occur to any company anywhere in the world. Or perhaps it was in direct response to a growing public awareness and demand for increased environmental protection, caused by the perception that environmental deterioration was now a *personal health issue*.

Whatever the reason, corporate America began to get its environmental house in order. During the 70s and early 80s, this "greening of the corporation" was a quiet one. It happened without much external fanfare, without much public attention. It occurred through the development of corporate safety, health, and environmental policies and guidelines; the formation and beefing up of safety, health, and environmental staff groups; the allocation of considerable capital for improved environmental performance; and the integration of EH&S issues into the business planning process. Not all of industry, however, moved to this state in a uniform fashion. Farsighted CEOs saw this as a serious business issue and led the effort, while for others the environmental alarm clock sounded much later.

If the corporate greening had its roots in the 1970s, it began flourishing with the passage of the SARA Title III Amendments in November, 1996. For the first time, many industries were required to publicly report their emissions to air, water, and land. For some corporations, this was the first time that senior executives had seen the roll-up of these numbers. They were large, very large in some cases. Shock, disbelief, and embarrassment were some of the reactions, and out of this came a resolve to improve their corporation's environmental standing. Recognizing that external reporting of this data would produce similar reactions in the general public, companies felt it vital to tell the public what they were doing to reduce emissions and wastes. Goals for reduction were set and communicated to the public almost with a spirit of competition between companies, as each sought the environmental high ground.

At about the same time, market research findings showed that almost 25 percent of American shoppers were making purchases based on environmental decisions, and that, by 1995, greater than 50 percent of the population would become "green" purchasers. If the consumer then was to base his or her purchases on the environmental aspects of a product, it was only a small mental leap for the consumer to buy products only from companies with a good environmental reputation. The major players in the chemical and other "environmental impact" industries responded by launching major environmental initiatives made public with green advertising. The realization that environmental protection and the bottom line were not mutually exclusive, led to the next realization that environmental responsibility and leadership could yield competitive advantage. To be recognized as a *benchmark* company in environmental stewardship took on an importance that was not trivial.

During this same twenty-year period of first quiet, then public "greening", senior management, having watched and studied the factors which led to Japan's meteoric rise in worldwide economic stature, began to pursue the same success factors: total quality management, organizational effectiveness and functional excellence, and continuous improvement.

## BENCHMARKING AS A TOOL

One of the techniques or tools to help set targets for continuous improvement is the practice of *benchmarking*. David T. Kearns, former Chief Executive Officer of Xerox Corporation, defined benchmarking as "the continuous process of measuring products, services, and practices against the toughest competitors or those companies recognized as industry leaders." It's a process of rigorously measuring your performance versus the "best in class" companies and using the information obtained to establish goals, targets, priorities, and practices to meet or surpass the best in class and thereby gain competitive advantage.

The technique is well suited to the EH&S function as a means to search for best practices that will lead to superior performance. In other words, it allows a company to determine how their approach and results stack up relative to other companies and it can help them to find best practices that can improve performance, or to find lower cost alternatives at the same performance level.

Benchmarking, in its traditional form, is a relatively simple process in concept, but it requires a good deal of up-front planning to maximize its benefits. There must also be the organizational willingness to make change based on the results of the benchmarking, otherwise an expensive, time consuming benchmarking activity will produce little benefit.

## BENCHMARKING PHASES—TRADITIONAL

Benchmarking is composed of a planning phase, a data collection phase, an analysis phase, and an integration and implementation phase. (See Figure 1, page 291.)

### Planning Phase

The planning phase is critical, because this is where decisions are made on the scope of the benchmarking effort. In the EH&S arena, the scope could be as broad as benchmarking the total EH&S effort of your company, or more focused, such as a company's EH&S strategic planning processes, or how a company manages EH&S information, or how it approaches process safety management, or how it manages its EH&S training. I have personally organized and resourced benchmarking efforts that have looked at organization and staffing issues and where certain bodies of work are performed within organizations for maximum effectiveness; how various types of remediation activities are managed; and how various aspects of audit programs are organized and managed. Not only individual companies have utilized benchmarking studies for continuous improvement, but industry groups have as well. For example, the Business Roundtable sponsored a major cross-industry, facility-level benchmarking effort on pollution prevention and published the results in November 1993.

A good first cut on the selection of areas of your EH&S program that should be considered for benchmarking can be obtained by asking the following questions:

- What elements of the EH&S program are most critical for the success of the overall effort'?
- What areas are causing the most problems or are falling short of performance expectations?

Answers to these questions can yield areas where a benchmarking effort may be able to provide significant improvement.

After selecting the EH&S area(s) for benchmarking, the question needs to be asked and answered as to what the products of the benchmarking effort will be. Can these products be used later to set goals and develop action plans to attain superior performance and will they be informative enough to establish best practices and effective management systems. The focus of the next step, information or data collection, must then provide the products desired from the benchmarking effort. It is important to include operational measures of performance as part of the products to be delivered, as this will be important in the implementation and monitoring of the new processes and systems that will be put in place as a result of the benchmarking effort.

The next step is to identify comparative companies and against how many companies you will benchmark. Who are the best of the best in EH&S efforts? Certainly many large corporations today have mature EH&S programs that have been highlighted over the last several years in the media. But there are many mid- and smaller- size companies with excellent EH&S programs. A lot of benchmarking is done by selecting direct competitors within the industry, but the most productive benchmarking occurs when target companies are world class leadership companies, regardless of the business. In fact, looking at companies outside your industry can help break paradigms of what can truly be accomplished without being bound by the rules of the industry. The target companies selected should also be those that are future looking and are anticipating EH&S issues well into the next decade. Information on selecting target companies can be found in articles in national magazines, trade publications, public databases, and other generally available public sources. Consultants, with their broad client knowledge base, can also be of value in this phase and other phases of the process.

## Data Collection Phase

The next phase of the benchmarking process involves deciding the data that is to be collected and the data collection method. What is needed is good quality data of sufficient consistency and quantity to validate comparisons between the benchmarked companies. Data collection is time consuming, costly, and may require specialized expertise to obtain it and determine its quality. Some data can be obtained publicly or through carefully structured questionnaires, while other data can only be obtained through interviews and facility visits. In the interview process, consideration should be given to obtaining information from different parts and levels of the organization. What you learn from an EH&S vice-president about a company's EH&S efforts will be different than what you learn from an environmental professional. Also interviews with the provider of a service, such as regulatory interpretation and guidance, may yield different information than an interview with a recipient of that service, such as the facility environmental or safety professional. There is no one right way to obtaining benchmarking data because each benchmarking process is unique.

What information is collected, the detail involved, the comprehensiveness to provide a certain quality level, the cost, time, and effort it will take, and whether the data will provide meaningful information toward achieving the products of the benchmarking effort, will all be issues that need addressing in the data collection phase.

## Analysis and Comparison Phase

Once the data is obtained, the next step is to analyze it and compare it with your internal operations. I believe this analysis needs to be done with a cross section of the organization, professionals as well as managers, as well as with the participation of business and operating personnel. This will provide several viewpoints for the data analysis. Consultants can also be used here to obtain an independent analysis.

You perhaps will see some results that show that your benchmark element is at parity with the other benchmark companies and may decide not to pursue any further activity. However, parity may be short lived in today's competitive world, but also improvement in the parity area may lead to competitive advantage. If the benchmark companies have been properly selected, most of the results will identify "performance gaps," between your practices and performance and those of the best-in-class benchmark companies.

It is important at this stage to also project future performance levels of the best-in-class companies so that the performance gap is not measured in today's terms only. These best-in-class companies will not likely be standing still and, most likely, will have strategic EH&S plans to take them to even higher performance. Also necessary is to project the regulatory and public opinion trends that will impact your specific operations during the implementation phase so that they can be taken into account in developing your specific strategies and actions.

## Implementation Phase

Communication of the benchmarking analysis to the organization, particularly to those management personnel that have a direct impact on the function that is being benchmarked is critical for the next phase, which is to set goals and action plans—both strategic and tactical—to overcome the performance gap or rise above the performance gap. Without leadership recognition of a performance gap and the support to close it, a benchmarking study is an expensive exercise with minimal payback. The communication needs to be well thought out, as there is always some skepticism that arises whenever new practices are to be implemented. The communication by the benchmarking team needs to occur throughout the progress of the benchmarking study so that questions from management, field personnel, and the line can be addressed during the course of the benchmarking rather than when most of the effort has been completed. This will avoid the "surprises" that can delay or even destroy the implementation process.

Organization of the analysis so that is it credible and understandable for each particular audience that you will be addressing is also important.

Of course the setting of goals and action plans and the implementation of those plans, monitoring of progress toward closing the performance gap, and future recalibrations are also included in a dynamic benchmarking process. Again, the benchmarking team should engage as many members of the organization as practical to assist in the goalsetting process and the development of action plans. The action plans should address what task will be done, in what sequence, by whom, and when it will be accomplished. Management support will be needed to allocate the appropriate resources to assure successful completion of the action steps necessary to overcome the "performance gaps" that were identified.

Benchmarking is a continuous evergreen process, not an end product or a one-shot, one-time analysis. Therefore, a schedule should be developed as to when the best practices and resultant performance will again be reviewed opposite the best-in-class. This recalibration should be a shorter, less time consuming, less costly effort than the original benchmarking since the process will be familiar and the scope will be more defined.

These are the basic elements of a formal benchmarking process. There is certainly more that has been written on the topic, and in significantly greater detail, but this outline should prove useful in considering benchmarking as part of your overall EH&S management systems for continuous improvement.

## *Figure 1*

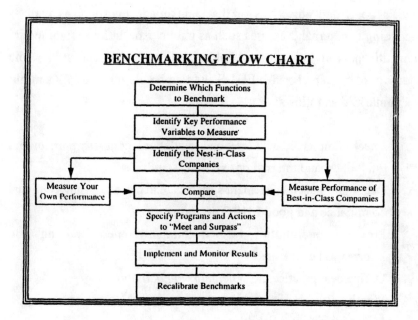

**BENCHMARKING FLOW CHART**

Determine Which Functions to Benchmark

Identify Key Performance Variables to Measure

Identify the Nest-in-Class Companies

Measure Your Own Performance

Compare

Measure Performance of Best-in-Class Compamies

Specify Programs and Actions to "Meet and Surpass"

Implement and Monitor Results

Recalibrate Benchmarks

## LESS FORMAL BENCHMARKING PROCESSES

There are also less formal processes that have some of the core benchmarking attributes that can be used to improved EH&S performance. Two are discussed below.

### Auditing

The role of environmental auditing has changed significantly since it's inception in the early 1970s, having matured from being merely a mechanism for evaluating regulatory compliance to now being perceived as an invaluable element in a proactive EH&S strategy. The approach to, and process of auditing has to be an evolving one, designed to address the specific goals and objectives of the company in question, as well as responding to external pressures such as public expectations, commercial competitiveness, investor and stockholder confidence, risk, and liability. The objectives of a comprehensive EH&S auditing program in the 1990s might be summarized as follows:

- Check compliance with laws and regulations, company policies and guidelines, and internal or external "standards;"
- Determine what management systems are in place to ensure continual compliance and good EH&S performance;
- Provide opportunities for education/awareness training and information exchange;
- Define best practices; *[benchmarking activity]*
- Identify areas for improvement; *[benchmarking activity]*
- Minimize risk and liability;
- Identify and prioritize issues for focused attention;
- Optimize resource allocation.

From the auditing process, an informal benchmarking tool is available for companies to find best practices within their own organizations and also areas for improvement For the auditing process to achieve these ambitious goals, a different process than that for merely assessing regulatory compliance is needed. There has to be a real and sustained commitment by management and active involvement and support by the operations and field personnel to make this process work. Training of audit personnel needs to be more rigorous, and a process for sharing best practices must be in place. See Chapter 14 for a more comprehensive discussion on auditing.

## Cooperative Benchmarking

I have resourced several benchmarking efforts for Pilko & Associates, Inc. (Pilko) utilizing a less formal approach which Pilko has termed Cooperative Benchmarking™. This involves a process where a number of corporations (usually seven or eight), with similar EH&S concerns, join together to candidly discuss pre-determined elements of their EH&S programs. They share the costs and benefits of the benchmarking effort equally. A major advantage is that the costs for such an effort will be significantly less than if a single company commissioned a study by outside consultants.

A scope for the effort is determined along with the particular EH&S elements to be benchmarked (planning phase). A detailed questionnaire is prepared probing aspects of the benchmark elements and is filled out by each participating company, with all of the companies receiving each other's completed questionnaires. The questionnaire provides background and the baseline for the remainder of the process. Then a series of face-to-face two-day meetings is held with three or four companies at a time until all companies have had the opportunity to meet face-to-face with each other.

These meetings, facilitated by the consultant, result in a significant amount of information sharing and in depth discussion between the participating companies. Unique approaches as well as "best practices" evolve from the discussion. This completes the data collection phase.

A detailed report is then prepared which summarizes the discussion points, highlights unique approaches, and compares benchmark data among the companies (analysis phase). Companies participating in the process appreciate the candid face-to-face dialogue and the insights they get into other companies' approaches in dealing with the same issues they face.

## BEST-IN-CLASS EH&S PROGRAMS

Throughout the preceding discussion, there has been reference to measuring your performance versus "best-in-class" companies in the EH&S arena. The following describes what might be the characteristics of a best-in-class EH&S program for a major corporation in the 1990s.

• There is a strong senior management commitment to a proactive EH&S agenda for the corporation. This means direct involvement in the EH&S objective-setting process as well as providing resource support for environmental activities. EH&S topics are a regularly scheduled agenda item for senior management discussion sessions, and a board committee, which includes outside directors, oversees the corporation's efforts in the EH&S arena. The CEO views him/herself as the company's chief EH&S officer, rather than yielding total responsibility to the EH&S vice-president. The annual report contains information to stockholders on the EH&S performance of the corporation. A separate, more detailed, report on the company's EH&S performance is also issued annually.

- A corporate EH&S mission statement or policy exists that clearly communicates that EH&S responsibility is a core corporate value and an integral part of doing business. This has been communicated throughout the company so that all levels recognize it as a part of doing their job. Appropriate guidelines have been developed on significant EH&S topics to achieve a consistent business approach on these issues. Short- and long- term EH&S goals are set and progress is monitored.

- Standards exist that govern operations worldwide, even where national or local laws and regulations may be sparse or unenforced. These standards offer a safe margin of protection, with particular emphasis on environmentally sensitive areas.

- EH&S performance is a line management responsibility and EH&S staff groups act as resources in assisting the line to discharge their responsibilities. Compensation is tied directly to environmental performance and those that excel are rewarded and recognized for their efforts.

- There is a corporate EH&S staff group, which is headed by a senior level manager who has direct access to the executive leadership. It's role and responsibilities are clearly defined. This group works closely with legal, public affairs, governmental affairs, engineering and R&D to guide and support the line's EH&S efforts. Other EH&S functional groups exist, as appropriate, at departmental or division and facility levels.

- An EH&S strategic planning process is in place to assure that EH&S issues—past, present, and future—are identified so that strategies, action plans, and resource needs are deployed proportionate to the risk they pose. This is an integral part of the business planning process.

- The individual businesses understand the magnitude of the impact that their processes and products have on the environment and have a strong pollution prevention bias.

- There is active participation in the statutory and regulatory process—either through trade associates or direct individual involvement—to bring about laws and regulations that are technically sound, produce tangible results, and are reasonable from a cost/benefit standpoint. This is done in a nonadversarial, constructive fashion. Compliance with the letter and spirit of laws and regulations is a minimum requirement. Initiatives beyond regulatory requirements are undertaken when the added benefit is reasonable relative to the cost.

- Public input is sought to assist in shaping EH&S programs. This takes the form of citizen advisory councils at the local or regional level and developing partnerships with conservation and safety organizations at the corporate level. Participation in environmental and safety education efforts at the grammar and high school levels and sponsorship of research fellowships at colleges and universities are other examples of proactive approaches.

- Environmental, health, and safety training takes place at all levels of the organization, assuring proficiency in the EH&S function and awareness and sensitivity within the business function. The EH&S function is populated with high performing individuals whose contributions are recognized as critical to business success.

- A prudent effort is in place to seek out environmental problems caused by past practices so that they can be mitigated if they pose an unreasonable risk. New facilities are constructed, based on pollution prevention and process safety design criteria. The EH&S function is a member of all major project teams.

- The R&D organization develops products and processes with environmental protection, process safety, and product safety as key considerations. The product stewardship activity resides in the business units.

- A well-designed EH&S quality assurance program is in place which audits facilities for compliance to applicable laws, regulations, and corporate policies. In addition, it assures that management systems are in place to perpetuate compliance and manage risk.

- A system exists to analyze all acquisitions and divestitures for their EH&S risk prior to approval.

- The continuous improvement and total quality management process are utilized in the EH&S framework to maintain leading edge performance. Best practices are shared broadly across the company. Benchmarking is one tool used to assess EH&S leadership and promote continuous improvement.

It is evident from the foregoing that a best-in-class EH&S program is a multi-faceted effort that touches on nearly all functions and individuals in the company. No one company may be best-in-class in all activities. Several are striving to be. These companies take pride in what they have accomplished, while still recognizing there is more to do to continue to meet public expectations and to assure that they are meeting their regulatory obligations. They are usually willing to share successes and, in some cases, their failures. Benchmarking is a way to do this.

# APPENDIX

## INTERNATIONAL STANDARDS ORGANIZATION (ISO) STANDARD 14000 (DRAFT) — ENVIRONMENTAL MANAGEMENT

*Gordon A. West, Pilko & Associates, Inc.*

*Steven R. Cumbow, and Kristine F. Link, Arthur Andersen & Company*

References are made throughout the chapters of this book to the pending ISO Standard 14000 on environmental management. This standard is not final at the time of publication of this volume. The first section of this standard is in final draft form and 14001, covering environmental systems, is expected to be adopted in 1996. Leading corporations with global operations are planning and making changes in their programs to enable them to be in compliance with the standard when or soon after it does become final. These actions are being taken because ISO 14001 is a common sense template for effective management and conformity with the standard may become a requirement for remaining competitive in world markets.

ISO 14001 will not be a regulatory compliance standard. It will identify the core elements of a management system that will enable a company to set objectives monitor and measure performance, and demonstrate what level of performance has been achieved. Conformity with ISO 14001 will require companies to

- secure management and employee commitment to the protection of the environment with clear assignment of accountability; and responsibility;
- determine the environmental aspects associated with the organization's activities, products and services and meet the legal requirements that govern them;
- encourage strategic planning to reduce environmental impact throughout the life cycle of all products;
- establish structured management processes to achieve targeted performance levels;
- provide the resources necessary to achieve targeted performance levels;
- evaluate and document performance against policies objectives and targets;
- and periodically review the environmental management system and identify opportunities for improvement.

ISO 14000 is intended to become a companion standard to ISO 9000 which, in a few short years, has become a worldwide standard for registration of total quality management (TQM) systems. ISO 9000 provides the basis for independent auditors to certify that a facility has met TQM requirements and provides proof to customers that a supplier has systems in place to provide quality products and services. While ISO 9000 is focused on customer interests, ISO 14000 will address the interests and concerns of all stakeholders. Despite this difference, there are many similarities in the requirements of both standards and the approach taken to secure registration under one will be helpful in obtaining registration under
the other. The following chart compares the provisions of ISO 14001 (Draft) and ISO 9001.

# COMPARISON OF ISO 14001 AND ISO 9001

ISO 14001

4.1 -- *Environmental Policy*
Top management shall define the
organization's environmental policy
and ensure that it;

(a) is *appropriate* to the nature,
scale and environmental impacts of
its activities, products and services;

(b) includes a *commitment to*
*continual improvement* and
prevention of pollution;

(c)includes a *commitment to*
*comply with relevant*
*environmental legislation and*
*regulations*, and with other
requirements to which the
organization subscribes;

(d) provides the *framework for*
*setting and reviewing*
*environmental objectives and*
*targets*;

(e) is *documented, implemented,*
*maintained, and communicated* to
all employees; and

ISO 9001

4.4.1 -- *Quality Policy*
The supplier's management with
executive responsibility shall *define*
and document its policy for
quality....
The quality policy shall be *relevant*
to the supplier's organizational
goals and the expectations and
needs of its customers.
...defines and documents its policy
for quality, including its
*commitment to quality.*

Not included.

Not included.

The supplier's management with
executive responsibility shall define
and *document its policy* for

understood, implemented, and

quality....The supplier shall ensure that this policy is *understood, implemented, and maintained* at all levels of the organization.

(f) is available to the public;

Not included.

4.2 -- *Planning*

4.2.3 -- *Quality Planning*

4.2.1 -- *Environmental Aspects*

The organization shall establish and maintain a procedure to identify the environmental aspects of its activities, products, or services that it can control and over which it can be expected to have an influence, in order to determine those which have or can have significant impacts on the environment. The organization shall ensure that these impacts are considered in setting its environmental objectives;

The organization shall keep this information up-to-date.

4.2.2 -- Legal and Other Requirements.

The organization shall establish and

4.4.4 -- Design input requirements relating to the product, including maintain a procedure to identify and have access to *legal, and other requirements* to which the organization subscribes directly applicable to the environmental aspects of its activities, products or services.

applicable *statutory and regulatory requirements*, shall be identified, documented, and their selection reviewed by the supplier for adequacy.

4.2.3 -- Objectives and Targets

The organization shall establish and maintain documented environmental *objectives and targets*, at each relevant function and level within the organization.

4.1.1 -- Quality Policy: define and document is policy for quality, including objectives for quality....

4.2.3(g) -- Quality Planning: the clarification of standards of acceptability for all features and requirements, including those which contain a subjective element.

When establishing and reviewing its objectives, an organization shall consider the legal and other requirements, its significant environmental aspects, its technological options, and its financial, operational and business requirements, and the views of interested parties.

Not included.

The objectives and targets shall be consistent with the environmental policy, including the commitment

Not included.

To prevention of pollution.

4.2.4 -- Environmental
Management Programme(s)

The organization shall establish and
maintain (a) programme(s) for
achieving its *objectives and targets*.
It shall include:

(a) designation of *responsibility for
achieving objectives and targets* at
each relevant function and level of
the organization, and

(b) the *means and time frame* by
which they are to be achieved.

If appropriate, programme(s) shall
be amended to ensure that
environmental management will
also apply to projects relating to
*new developments and new or
modified activities, products or
services*.

4.3 -- Implementation and
Operation

4.3.1 -- Structure and
Responsibility

4.1.2.1 -- The *responsibility*,
authority, and the interrelation of
personnel who manage, perform
and verity work affecting quality
shall be defined and documented....
Not included.

Not included.

*Roles, responsibilities, and authorities* shall be defined, documented, and communicated in order to facilitate effective environmental management.

Management *shall provide resources* essential to the implementation and control of the environmental management system. Resources include human resources and specialized skills, technology, and financial resources.

The organization's top management *shall appoint (a) specific management representative(s)* who, irrespective of other responsibilities, shall have defined roles, responsibilities and authority for:

(a) *ensuring that environmental management system requirements are established, implemented and maintained in accordance with this standard*;

(b) *reporting on the performance* of the environmental management system to top management for

4.1.2.2 -- The *responsibility, authority and the interrelation* of personnel who manage, perform, and verify work affecting quality shall be defined and documented....

4.1.2.2 -- The supplier *shall identify resource requirements and provide adequate resources....*
4.1.2.3 -- The supplier's management with executive responsibility *shall appoint a member of the supplier's own management* who, irrespective of other responsibilities, shall have defined authority for (a) ensuring that a quality system is established, implemented, and maintained in accordance with the International Standard, and

(b) *reporting on the performance* of the quality system to the supplier's management for review and as a review and as a basis for improvement of the environmental management system.

basis for improvement of the quality system.

4.3.2 -- Training, Awareness, and Competence.

4.18 -- Training

The organization shall *identify training needs. It shall require that all personnel whose work may create a significant impact upon the environment, have received appropriate training.*

The supplier shall **establish and maintain documented procedures for identifying training needs and provide for the training** of all personnel performing activities affecting quality.

Is shall establish and maintain procedures to make its employees or members at each relevant function and level aware of

(a) the importance of conformance with the environmental policy and procedures and with the requirements of the environmental management system;

Not included.

(b) the significant environmental impacts, actual or potential, of their work activities and the environmental benefits of improved personal performance;

Not included.

(c) their roles and responsibilities in achieving conformance with the environmental policy and procedures, and with the requirements of the environmental management system, including emergency preparedness and response requirements; and

(d) the potential consequences of departure from specified operating procedures.

*Personnel performing the tasks which can cause significant environmental impacts shall be competent on the basis of education, appropriate training and/or experience, as required.*

### 4.3.3 -- *Communications*

The organization shall establish and maintain procedures for: (a) *internal communication* between the various levels and functions of the organization; and

(b) receiving, documenting, and responding to relevant communication from external

*Personnel performing specific assigned tasks shall be qualified on the basis of appropriate education, training, and/or experience, as required.* Appropriate records of training shall be maintained.

Not included.

Not included.

interested parties regarding environmental aspects and the environmental management system.

The organization shall consider *processes for external communication* on significant environmental aspects and record its decision.

### 4.3.4 -- *Environmental Documentation*

The organization shall establish and maintain information, in paper or electronic form, to

(a) describe the core elements of the management system, and their interaction, and

(b) provide direction to related documentation.

### 4.3.5 -- *Document Control*

The organization shall establish and maintain procedures for controlling all documents required by this standard to ensure that

4.2.1 -- The supplier shall establish, *document*, and maintain a quality system....shall prepare a quality manual covering the requirements of this International Standard. The quality manual shall include or make reference to the quality system procedures, and outline the structure of the documentation used in the quality system.

### 4.5 -- *Document and Data Control*

4.2.3 -- Quality Planning: The supplier shall define and *document* how the requirements for quality will be met.

(a) they can be located;

(b) they are periodically reviewed, revised as necessary, and approved for adequacy by authorized personnel;

(c)the current versions of relevant documents are available at all locations where operations essential to the effective functioning of the system are performed;

(d) obsolete documents are promptly removed from all points of issue and points of use, or otherwise assured against unintended use; and

(e) any obsolete documents retained for legal and/or knowledge preservation purposes are suitably identified.

Documentation shall be legible, dated (with dates of revision) and readily identifiable, maintained in an orderly manner, and retained for a specific period. Procedures and responsibilities shall be established and maintained concerning the

creation and modification of the
various types of document.

4.3 -- Contract Review. The
supplier will establish and maintain
*documented* procedures for contract
review and for the coordination of
these activities.

4.4 -- Design Control. Design input
requirements relating to the
product, including applicable
statutory and regulatory
requirements, will be identified,
*documented*..

4.9 -- Process Control. The supplier
will identify and plan the
production, installation, and
servicing processes which directly
affect quality and shall ensure that
these processes are carried out
under controlled conditions.
Controlled conditions shall
include...
(a) documented procedures
defining the manner of production,
installation, and servicing, where
the absence of such procedures
could adversely affect quality.

4.15 -- Handling, Storage,

Packaging, Preservation, and Delivery. The supplier will establish and maintain documented procedures for handling, storage, packaging, preservation, and delivery of product.

4.16 -- Servicing. Where servicing is a specified requirement, the supplier shall establish and maintain documented procedures for performing, verifying and reporting that the servicing meets the specified requirements.

4.6.1 -- Purchasing. The suppler will establish and maintain *documented procedures* to ensure that purchased product conforms to specified requirements.

4.6.2 -- Evaluation of Subcontractors

4.6.3 -- Purchasing Data

4.6.4 -- Verification of Purchased Product.

4.7 -- Control of Customer-Supplied Product. The supplier

shall establish and maintain *documented procedures* for the control of verification, storage, and maintenance of customer-supplied product provided for incorporation into the supplies or for related activities.

4.8 -- Product Identification and Tractability. Where appropriate, the supplier will establish and maintain *documented procedures* for identifying product by suitable means from receipt and during all stages of production, delivery, and installation.

### 4.3.6 -- *Operational Control*

The organization shall identify those operations and activities that are associated with the identified significant environmental impacts and which fall within the scope of its policy, objectives, and targets. The organization will plan these activities, including maintenance, in order to ensure that they are carried out under specified conditions by:
(a) establishing an maintaining documented procedures to cover situations where their absence could lead to deviations from the environmental policy, and the objectives and targets;
(b) stipulating operating criteria in the procedures;
(c)establishing and maintaining procedures related to the identifiable significant environmental aspects of goods and services used by the organization and communicating relevant procedures and requirements to suppliers and contractors.

### 4.3.7 --Emergency Preparedness and Response

4.2.3(g) -- Quality Planning...the clarification of *standards of acceptability* for all features and requirements, including those which contain a subjective element.

Not included.

The organization shall establish and
maintain procedures to identify
potential for and respond to
accidents and emergency situations,
and for preventing and mitigating
the environmental impacts that may
be associated with them.

The organization shall review and
revise, where necessary, its
emergency preparedness and
response procedures, in particular,
after such an occurrence of
accidents or emergency situations.

The organization shall also
periodically test such procedures
where practicable.

4.4 -- Checking and Corrective
Action

4.4.1 -- Monitoring and
Measurement

The organization shall establish and
maintain procedures to *monitor and
measure* on a regular basis the key
characteristics of its operations and
activities that can have a significant
impact on the environment. This
shall include the recording of
information on track performance,
relevant operational controls, and
conformance with the
organization's objectives and
targets

4.10 -- Inspection and Testing. The
supplier shall establish and
maintain documented procedures
for *inspection and testing* activities
in order to verify that the specified
requirements for the products are
met. The required inspection an
testing, and the records to be
established, will be detailed in the
quality plan or documented
procedures.

4.12 -- The inspection and test
status of product shall be identified
by suitable means, which indicate
the conformance or non-
conformance of product with regard
to inspection and tests
performed...throughout production,
installation, and servicing of
product.

4.20 The supplier shall identify the
need for statistical techniques
required for establishing,
controlling, and verifying process
capability and product
characteristics...documented

316 / Principles of Environmental, Health and Safety Management

Monitoring equipment shall be calibrated and maintained and records of this process shall be retained according to the organization's procedures.

The organization shall establish and maintain procedures for defining responsibility an authority, for handling and investigating non-conformance, taking action to mitigate any impacts caused by initiating corrective and preventive action.

4.4.2 -- Non-conformance and corrective and preventative action

procedures implement and control the application of the statistical techniques.

4.11 The supplier will establish and maintain documented procedures to control, calibrate, and maintain inspection, measuring, and test equipment (including test software), used by the supplier to demonstrate the conformance of product to specified requirements.

The organization will establish and maintain procedures for defining responsibility and authority, for *handling and investigating non-conformance, taking action to mitigate any impacts caused and for initiating corrective and preventive action.*

4.13 -- The supplier will establish and maintain documented procedure to ensure that *product that does not conform to specified requirements* is prevented from unintended use or installation. This control will provide for *identification, documentation, evaluation, segregation* (when practical), disposition of non-conforming product, and for notification to the functions concerned.

Any corrective or preventive action taken to eliminate the causes of actual and potential non-conformances shall be *appropriate to the magnitude of problems, and commensurate with the environmental impact encountered.*

The organization shall *implement and record* any changes in the documented procedures resulting from corrective and preventive action.

4.14 -- The supplier will establish and maintain documented procedures for *implementing corrective and preventive action to a degree appropriate to the magnitude of problems commensurate with the risks encountered implement and record* any changes.

4.4.3 -- *Records*

4.16 -- *Control of Quality Records*

The organization shall establish and

The supplier will establish and

maintain procedures for the *identification, maintenance, and disposition of environmental records*. These records will include training records and the results of audits and reviews.

*Environmental records will be legible, identifiable, and traceable* to the activity...product or service involved. Environmental records shall be *stored* and maintained in such a way that they are readily retrievable and protected against damage, deterioration, or loss. Their retention times will be established and recorded.

Records will be maintained, as appropriate to the system.

maintain documented procedures for *identification, collection, indexing, access, filing, storage, maintenance, and disposition* of quality records.

All *quality records will be legible*, and will be *stored* and retained in such a way they are readily retrievable in facilities that provide a suitable environment to prevent damage or deterioration and to prevent loss. Retention times of quality records will be established and recorded....Records may be in the form of any type of media....

and to the organization, to demonstrate conformance to the requirements of this standard.

### 4.4.4 -- *Environmental Management System Audit*

The organization will establish and maintain (a) programme(s) and procedures for periodic environmental management system audits to be carried out, in order to (a) determine whether or not the environmental management system

(1) *conforms to planned arrangements* for environmental management, including the requirements of this standard;

(2) has been *properly implemented and maintained*; and

(b) provide information on the results of the audit to management for its review.

### 4.17 -- *Internal Quality Audits*

The supplier will establish and maintain documented procedures for planning and implementing internal quality audits

to verify whether quality activities and related results,

*comply with planned arrangements.*

and to *determine the effectiveness* of the quality system.

The results of the audits will be recorded (see 4.16), and brought to the attention of the personnel having responsiblity in the area audited shall take timely corrective action on deficiencies found during the audit.

The audit program, including any *schedule, will be based on the environmental importance of the activity concerned and the results of previous audits.* In order to be comprehensive, the audit procedures will cover the audit scope, frequency, and methodologies, as well as the responsibilities and requirements for conducting audits and reporting results.

Internal quality audits will be *scheduled on the basis of the status and importance of the activity to be audited....*Follow-up audit activities will verify and record the implementation and effectiveness of the corrective action taken..

4.5 -- *Management Review*

The organization's management will, at intervals it determines, review the management system, to *ensure its continuing suitability, adequacy, and effectiveness.* The management review process will ensure that the necessary

4.1.3 -- *Management Review*

The supplier's management with executive responsibility will review the quality system at defined intervals sufficient to *ensure continuing suitability and effectiveness,* in satisfying the requirements of this information is collected to allow management to carry out this evaluation. This review will be documented.

The management review will address the possible need for changes to policy, objectives, and other elements of the environmental management system, in the light of environmental management system audit results, changing circumstances, and the commitment to continual improvement.

International Standard, and the supplier's stated quality policy and objectives. Records of such reviews will be maintained.

# INDEX

323

organizational and financial
commitment, 83, 97

manufacturing, 200

and marketing operations, 1

facilities, 10, 18, 197

processes, 22, 33, 74

Maria M. Bober, 133

material safety data sheets (MSDSs),
89, 106

material substitution, 197

matrix systems, 216

media relations, 60, 72, 148, 149

mergers, 226, 229

meteorology, 75

Ministry of Housing, 223

Mobil, 107

monitoring operations, 56, 57, 75, 194

multi-media program, 103, 190

National Enforcement Investigation
Center (NEIC), 255

national environmental policy plan,
214, 223

National Laboratories, 221

national parks, 230

National Pollutant Discharge
Elimination System (NPDES),
184

new products from wastes, 200

new technologies for pollution
prevention, 162

nitrous oxides, 28

non-E&S organizations, 60

non-nuclear energy efficiency, 203

nonhazardous waste, 188

nuisance and negligence, 90

Occupational Safety and Health
Administration (*see* OSHA)

off-site waste disposal liability, 248

Oil Pollution Act, 105

oil spill control, 274

operability, 230

operating and maintenance expense, 19

operating procedures, 168, 171, 172

organization, staffing, and structure,
247

organizational sentencing guidelines,
279

orientation for new employees, 83

OSHA, 87, 88, 99, 103

Hazard Communication Standard
Training, 88, 89

requirements, 88

standards, 88

oversight audit , 258, 259

oversight audits, 273, 275

oxidizing agents, 203

# GOVERNMENT INSTITUTES
# MINI-CATALOG

| PC # | ENVIRONMENTAL TITLES | Pub Date | Price |
|------|----------------------|----------|-------|
| 585 | Book of Lists for Regulated Hazardous Substances, 8th Edition | 1997 | $79 |
| 4088 | CFR Chemical Lists on CD ROM, 1997 Edition | 1997 | $125 |
| 4089 | Chemical Data for Workplace Sampling & Analysis, Single User | 1997 | $125 |
| 512 | Clean Water Handbook, 2nd Edition | 1996 | $89 |
| 581 | EH&S Auditing Made Easy | 1997 | $79 |
| 587 | E H & S CFR Training Requirements, 3rd Edition | 1997 | $89 |
| 4082 | EMMI-Envl Monitoring Methods Index for Windows-Network | 1997 | $537 |
| 4082 | EMMI-Envl Monitoring Methods Index for Windows-Single User | 1997 | $179 |
| 525 | Environmental Audits, 7th Edition | 1996 | $79 |
| 548 | Environmental Engineering and Science: An Introduction | 1997 | $79 |
| 578 | Environmental Guide to the Internet, 3rd Edition | 1997 | $59 |
| 560 | Environmental Law Handbook, 14th Edition | 1997 | $79 |
| 353 | Environmental Regulatory Glossary, 6th Edition | 1993 | $79 |
| 562 | Environmental Statutes, 1997 Edition | 1997 | $69 |
| 562 | Environmental Statutes Book/Disk Package, 1997 Edition | 1997 | $204 |
| 4060 | Environmental Statutes on Disk for Windows-Network | 1997 | $405 |
| 4060 | Environmental Statutes on Disk for Windows-Single User | 1997 | $135 |
| 570 | Environmentalism at the Crossroads | 1995 | $39 |
| 536 | ESAs Made Easy | 1996 | $59 |
| 515 | Industrial Environmental Management: A Practical Approa ch | 1996 | $79 |
| 4078 | IRIS Database-Network | 1997 | $1,485 |
| 4078 | IRIS Database-Single User | 1997 | $495 |
| 510 | ISO 14000: Understanding Environmental Standards | 1996 | $69 |
| 551 | ISO 14001: An Executive Repoert | 1996 | $55 |
| 518 | Lead Regulation Handbook | 1996 | $79 |
| 478 | Principles of EH&S Management | 1995 | $69 |
| 554 | Property Rights: Understanding Government Takings | 1997 | $79 |
| 582 | Recycling & Waste Mgmt Guide to the Internet | 1997 | $49 |
| 594 | Texas Environmental Regulations Manual | 1997 | $125 |
| 566 | TSCA Handbook, 3rd Edition | 1997 | $95 |
| 534 | Wetland Mitigation: Mitigation Banking and Other Strategies | 1997 | $75 |

| PC # | SAFETY AND HEALTH TITLES | Pub Date | Price |
|------|--------------------------|----------|-------|
| 547 | Construction Safety Handbook | 1996 | $79 |
| 553 | Cumulative Trauma Disorders | 1997 | $59 |
| 559 | Forklift Safety | 1997 | $65 |
| 539 | Fundamentals of Occupational Safety & Health | 1996 | $49 |
| 535 | Making Sense of OSHA Compliance | 1997 | $59 |
| 563 | Managing Change for Safety and Health Professionals | 1997 | $59 |
| 589 | Managing Fatigue in Transportation, *ATA Conference* | 1997 | $75 |
| 4086 | OSHA Technical Manual, Electronic Edition | 1997 | $99 |
| 598 | Project Mgmt for E H & S Professionals | 1997 | $59 |
| 552 | Safety & Health in Agriculture, Forestry and Fisheries | 1997 | $125 |
| 523 | Safety & Health on the Internet | 1996 | $39 |
| 597 | Safety Is A People Business | 1997 | $49 |
| 463 | Safety Made Easy | 1995 | $49 |
| 590 | Your Company Safety and Health Manual | 1997 | $79 |

▣ = Electronic Product available on CD-ROM or Floppy Disk

**GOVERNMENT INSTITUTES**

**G**

PUBLICATIONS CATALOG
1997

## PLEASE CALL OUR PUBLISHING DEPARTMENT AT (301) 921-2355 FOR A FREE PUBLICATIONS CATALOG.

### Government Institutes
4 Research Place, Suite 200 • Rockville, MD 20850-3226
Tel. (301) 921-2355 • FAX (301) 921-0373
E mail: giinfo@govinst.com • Internet: http://www.govinst.com

# G ORDER FORM

| Qty. | Product Code | Title | Price |
|------|-------------|-------|-------|
|  |  |  |  |
|  |  |  |  |
|  |  |  |  |
|  |  |  |  |
|  |  |  |  |
|  |  |  |  |
|  |  |  |  |

Subtotal_____

MD Residents add 5% Sales Tax_____

Shipping and Handling_____

Within US: Add $6/item for 1-4 items. Add $3/item for 5+ items/
Outside US. Add $15 /item for Airmail. Add $10/item for Surface

**Payment Enclosed**_____

Method of Payment

❏ Check (*payable to Government Institutes in US dollars*) $ _____

❏ Purchase Order (please attach to this order form)

❏ Credit Card:  Exp.___/____    ❏ MC   ❏ VISA   ❏ AMEX

Credit Card No. _____

Signature. _____

Name: _____
Company: _____
Address: _____
City:/ _____ State/Province: _____
Zip/Postal Code: _____ Country: _____
Telephone: _____ Fax: _____
E-mail Address: _____

## Government Institutes

4 Research Place, Suite 200 • Rockville, MD 20850-3226
Tel. (301) 921-2355 • FAX (301) 921-0373
E-mail: giinfo@govinst.com • Internet: http://www.govinst.com